RUSSIA BALLISTIC MISSILES

Updated 5 October 2022

Alexandre Zanfirov

DISCLAIMERS

The information and opinions contained in this document are provided "as is" and without any warranties or guarantees. Reference herein to any specific commercial products, process, or service by trade name, trademark, manufacturer, or otherwise does not constitute or imply its endorsement, recommendation, or favoring by the United States Government, and this guidance shall not be used for advertising or product endorsement purposes.

The statements of fact, opinion, or analysis expressed in this manuscript are those of the author and do not reflect the official policy or position of the Defense Intelligence Agency, the Department of Defense, or the U.S. Government. Review of the material does not imply DIA, DoD, or the U.S. Government endorsement of factual accuracy or opinion.

Copyright © 2020, 2021, 2022 - 4th Watch Publishing Co.

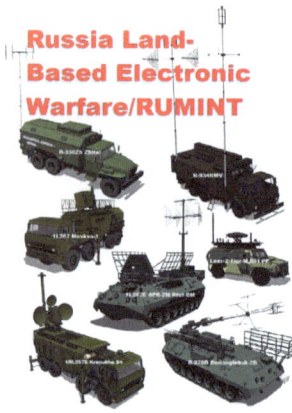 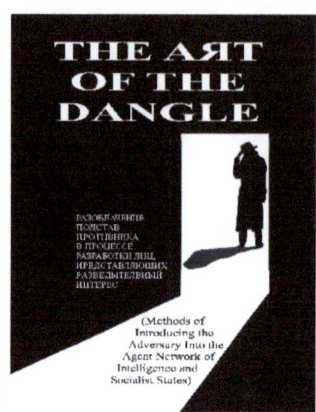  
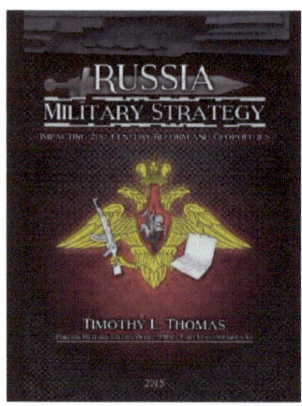  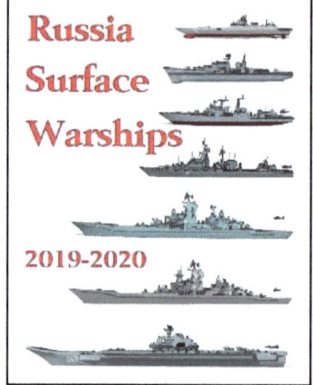 

 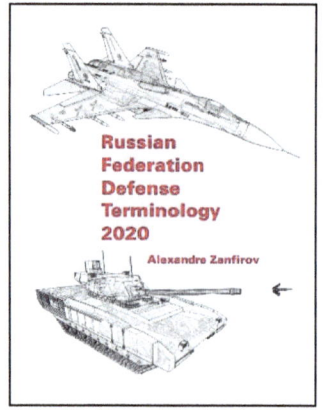  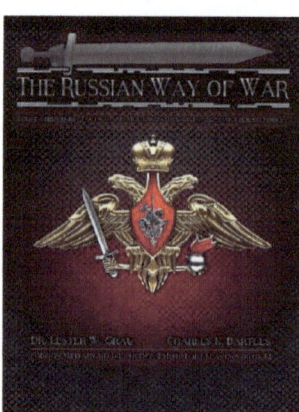

Other titles we publish on Amazon.com

Table of Contents

SRBM DECOY PENETRATION AID (PENAID) 1
HYPERSONIC MISSILES 3
Objekt 4202 ("Avangard" HGV) 5
SS-N-33 Zirkon (3M-22 Hypersonic Missile) 9
SA-N-9 Gauntlet (Kh-47M2 "Kinzhal" ALBM) 12
LAND-BASED BALLISTIC MISSILES 15
SS-18 Mod 5/6 Satan (R-36M2 "Voevoda" ICBM) 17
SS-19 Mod 4 Stiletto (UR-100N ICBM) 18
SS-21 Scarab (OTR-21 "Tochka" SRBM) 19
SS-24 Scalpel (RT-23 "Molodets" ICBM) 20
SS-25 Sickle (RT-2PM "Topol" ICBM) 21
SS-26 Stone (9K720 "Iskander" SRBM) 22
SS-27 Mod 1 Sickle B (RT-2PM2 "Topol-M" ICBM) 23
SS-29 (RS-24 "Yars" ICBM) 24
SS-X-30 Satan-2 (RS-28 "Sarmat" ICBM) 25
SS-X-31 Saber (RS-26 "Rubezh" ICBM) 26
SS-X-32Zh Scalpel B (RS-27 "Barguzin" ICBM) 27
SUBMARINE-LAUNCHED BALLISTIC MISSILES 29
SS-N-18 Stingray (R-29R "Vysota" SLBM) 31
SS-N-23 Skiff (R-29RMU2 "Layner" SLBM) 32
SS-NX-30 (RSM-56 "Bulava" SLBM) 33
CRUISE MISSILES 35
SSC-8 Screwdriver (9M729 GLCM) 37
SSC-X-9 Skyfall (9M730 "Petrel") 38
SS-N-19 Shipwreck (P-700 "Granit" ASCM) 39
SS-N-21 Sampson (RK-55 "Granat") 40
SS-N-25 Switchblade (Kh-35 "Uran") 41
SS-N-26 Strobile (P-800 "Oniks" ASCM) 42
SS-N-27 Sizzler (3M-54K "Kalibr" AShM) 43
SS-N-30A (Land Attack Cruise Missile) 44
AS-15 Kent (KH-55 ALCM) 45
AS-23A / AS-23B Kodiak (Kh-101/Kh-102 ASCM) 46
Russian Tactical Nuclear Weapons 49

Preface

Although most ballistic missiles are already hypersonic (defined as Mach 5 and above), not all of the missiles listed here can technically be called "Ballistic." The whole point of a ballistic missile is to deliver nuclear warhead(s) at great distances. Some ballistic missiles are ground-launched (in silos or mobile launchers, truck or train), and some are launched from submarines. A ballistic missile is one that once it has reached the apex of the flight path, the terminal trajectory is fairly predictable. Of course, assuming you think a ballistic missile will approach from the North pole, it's fair to expect an adversary to design a missile that goes the other direction and approaches from the South pole. Objekt 4202 (Avangard) is a hypersonic glide vehicle (HGV) specifically designed to avoid the flight path of traditional intercontinental ballistic missiles (to defeat anti-ballistic missile systems). The M-22 "Zirkon" Hypersonic Missile can be ground-launched, air-launched, or launched from a surface ship. While the speed of this missile is impressive, the fact is that once detected, in the amount of time a laser needs to reach the incoming missile it will have traveled about a quarter of an inch. It's not a question of can you intercept the missile as it is what part of the missile do you want to hit? Should the laser hit the warhead or the guidance and controls? I decided to also include cruise missiles of various sorts (GLCM, ALCM, ASCM and Hypersonic) because they too can be nuclear-capable.

March 16, 2022 - Added SRBM DECOY PENETRATION AID (PENAID)

I stand with Ukraine

DECOY PENETRATION AID (PENAID)

Penetration aids, or PENAIDs previously associated with ICBMs have been deployed by Russian 9K720 Iskander-M's solid-fuel 9M723 ballistic missiles in an attempt to spoof Ukrainian air defenses. Each missile carries 6 decoys that are roughly 16 inches long, dart-shaped, and with white bodies and an orange tail. According to intelligence officials, the decoys help the missiles evade air defense systems since they are able to "trick air defense radars and fool heat-seeking missiles." The tail fires flares to fool heat-seeking missiles.

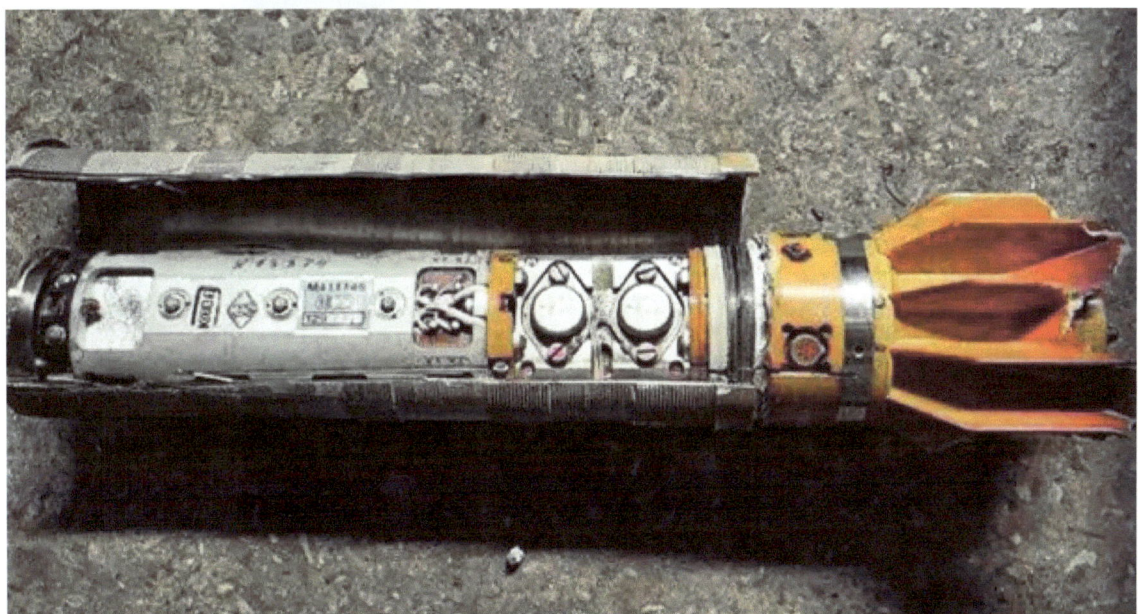

According to Michael Duitsman, the body's radar return adds clutter around the missile. The end of the PENAID has a flare to create a strong thermal signature. The interior of the PENAID has a radio transmitter to jam or spoof radars. The GRAU index is 9B899 and five different serial numbers have been reported, the highest being 2257. Two companies are known to be involved in the 9B899. The Central Radiotechnical Research Institute named for A.I. Berg issued contracts during the mid-2010s, and the Stavropol Radio Factory "Signal" claims to have started production of the 9B899 in 2008.

The intelligence official official speaking to the New York Times added that that the decoys are activated once targeted by air defense systems, suggesting that they may only be released once the Iskander-M missile determines that it is under threat.

HYPERSONIC MISSILES

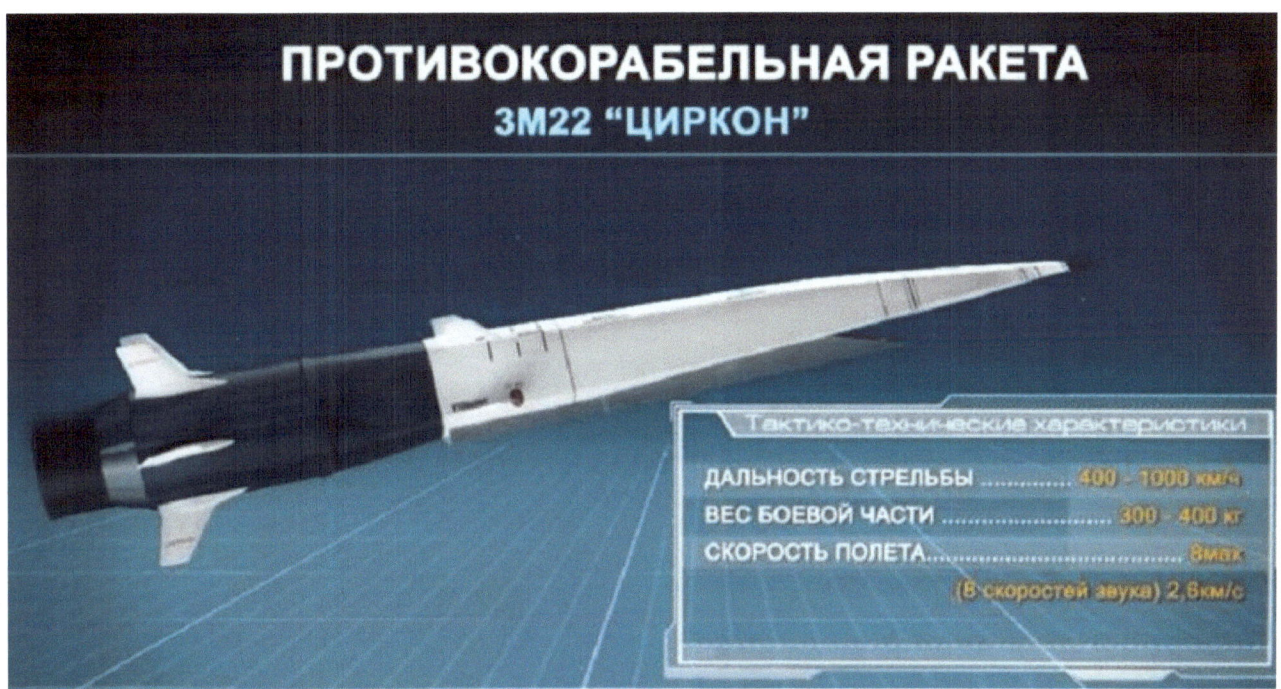

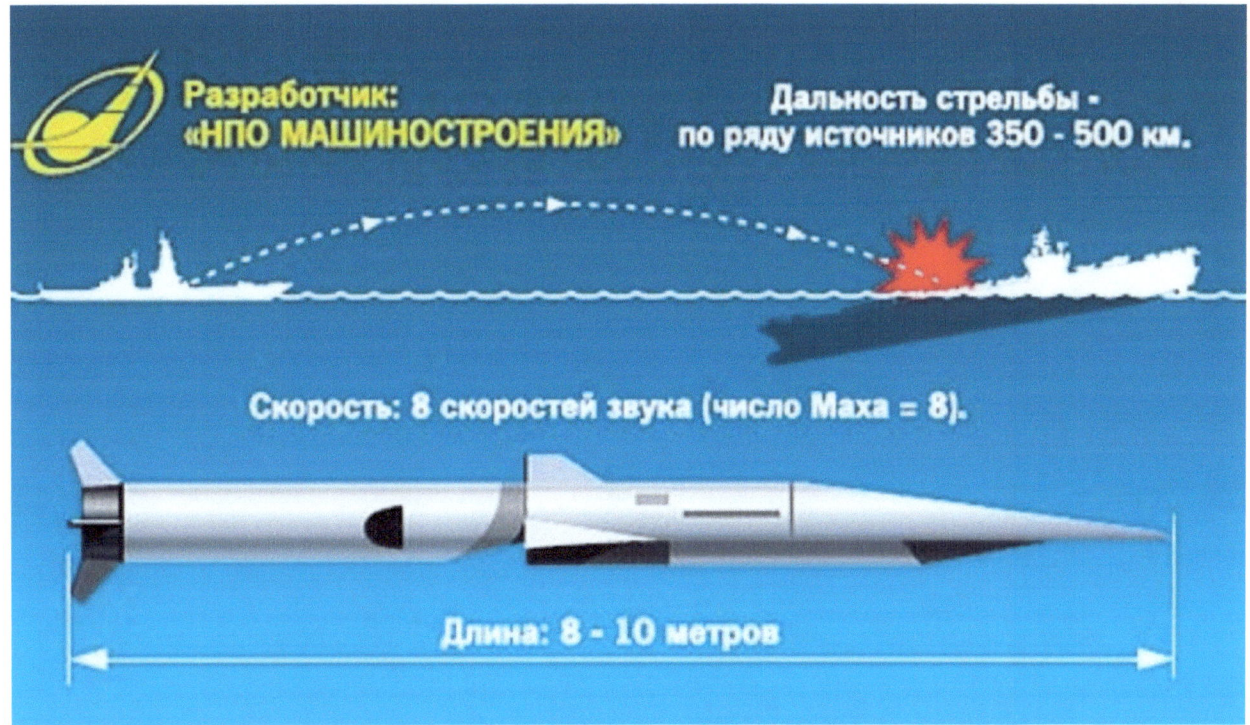

Yu-71

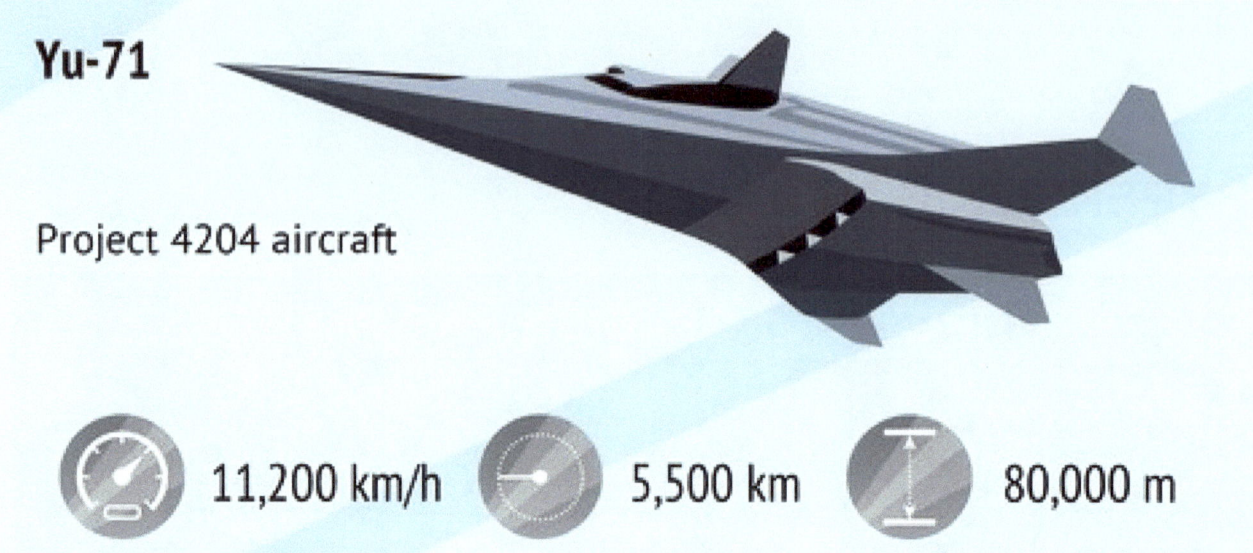

Project 4204 aircraft

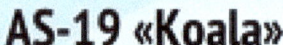

⏲ 11,200 km/h   🕐 5,500 km   ↕ 80,000 m

AS-19 «Koala»

Strategic hypersonic air-to-surface cruise missile

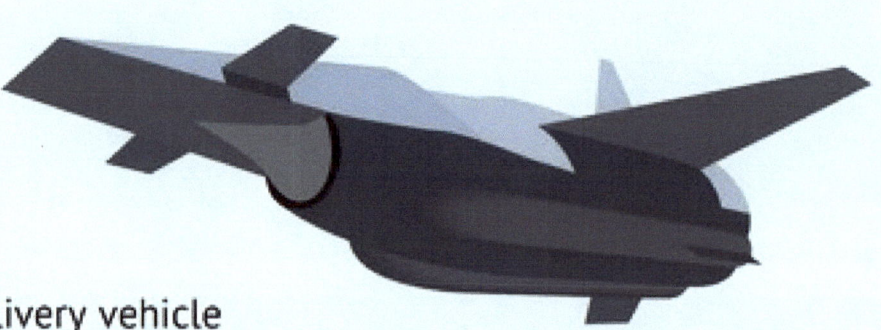

Tu-95 used as delivery vehicle

 5,310 km/h    3,000 km   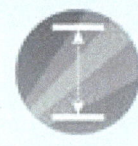 7,000 m (launch altitude)

Zircon 3M22

Sea-based hypersonic missile

50 times stronger kinetic energy at strike than existing air-to-ship or ship-to-ship missile. Will possibly enter service in 2018

 6,500 km/h    30,000 m

Objekt 4202 ("Avangard" HGV)

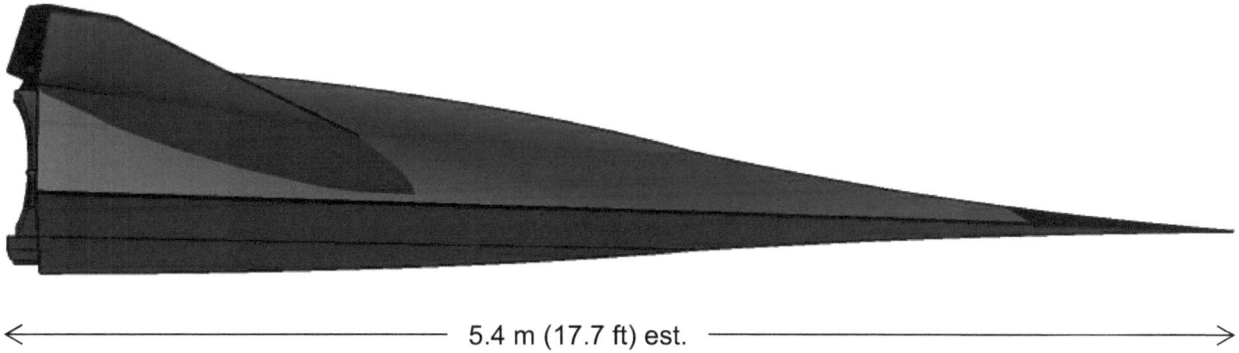

←——————— 5.4 m (17.7 ft) est. ———————→

The Objekt 4202 ("Avangard", Russian: Авангард; English: Vanguard), previously known as Yu-71 and Yu-74, is a Russian hypersonic glide vehicle (HGV), that can be carried as a MIRV payload by the UR-100UTTKh, R-36M2 and RS-28 Sarmat heavy ICBMs. (The RS-26 probably does not have the throw weight.) For now, it will be silo-based, but a mobile launcher is sure to follow (after 2027). Russia claims it can deliver both nuclear (2 MT) and conventional payloads at hypersonic speed of Mach 20 and higher from altitude of 80,000 km and a range of 5,500 km. Vehicle skin would need to be capable of enduring a temperature of 3500° Fahrenheit and extreme pressure for 3 minutes. On Dec. 26, 2018, the Avangard was test launched using an UR-100UTTKh ICBM. Putin has stated that the Avangard is capable of performing sharp maneuvers on its way to targets, making it "absolutely invulnerable for any missile defense system." Currently deployed with the 13th Red Banner Rocket Division at Yasny, Orenburg Oblast.

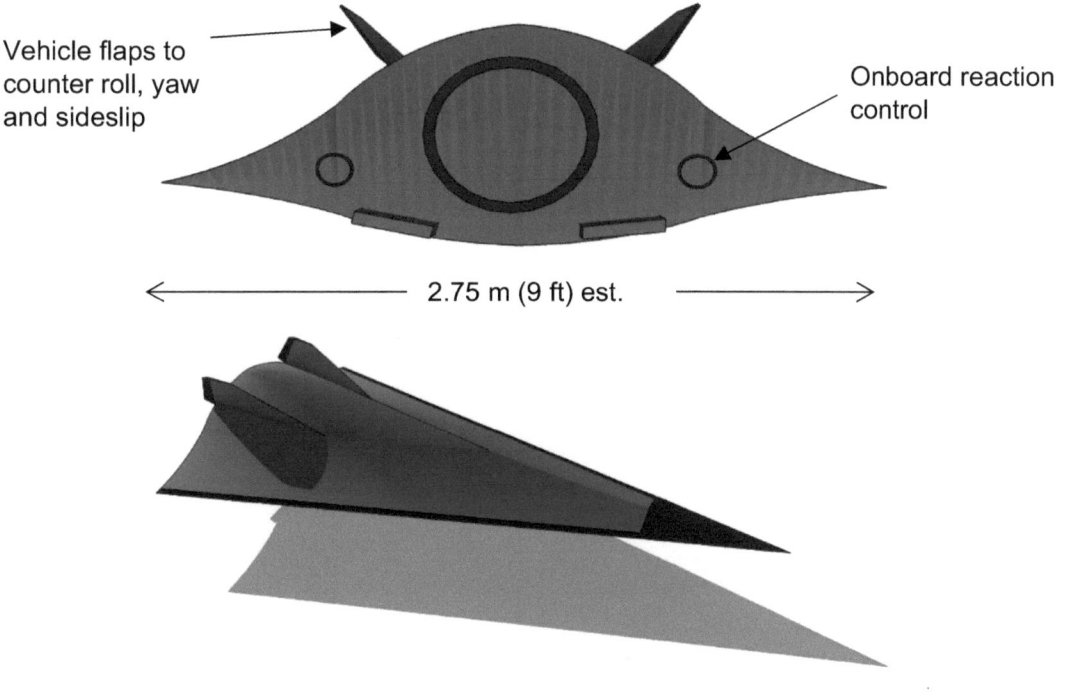

Vehicle flaps to counter roll, yaw and sideslip

Onboard reaction control

←——————— 2.75 m (9 ft) est. ———————→

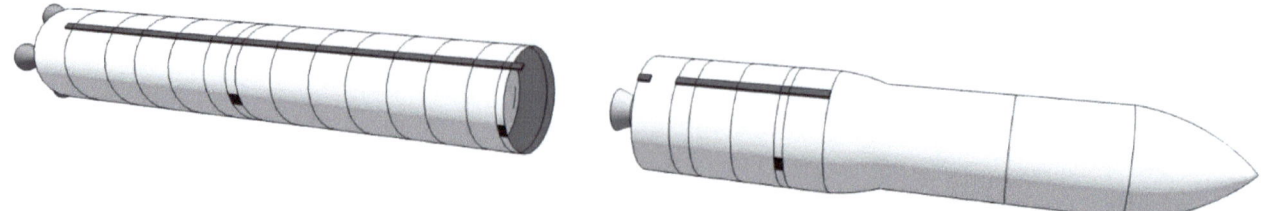

SS-18 post-boost vehicle separates from first stage

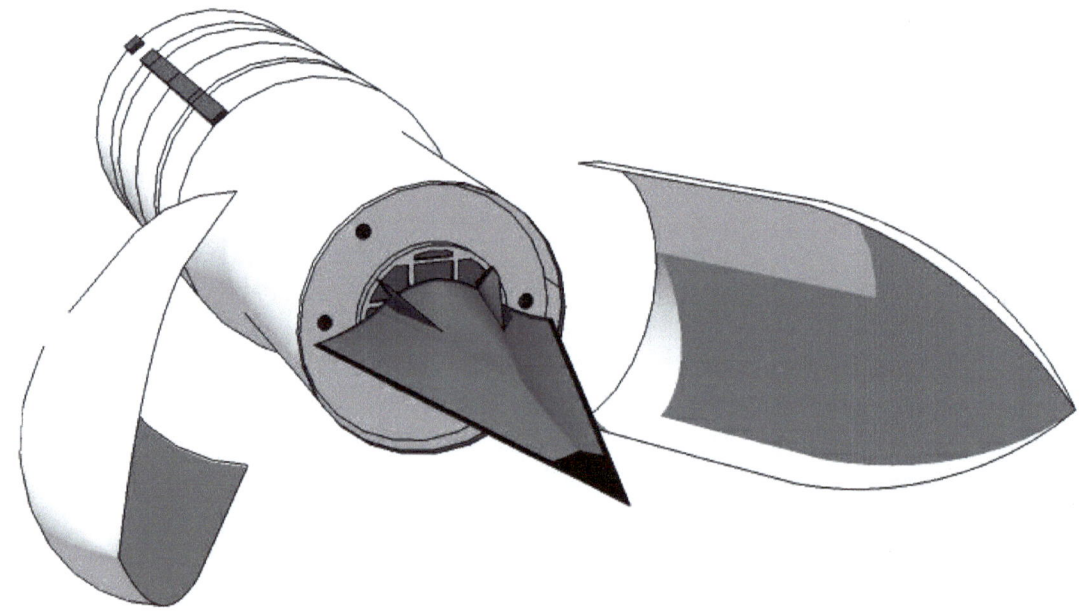

Post-boost vehicle jettisons nosecone exposing hypersonic glide vehicle

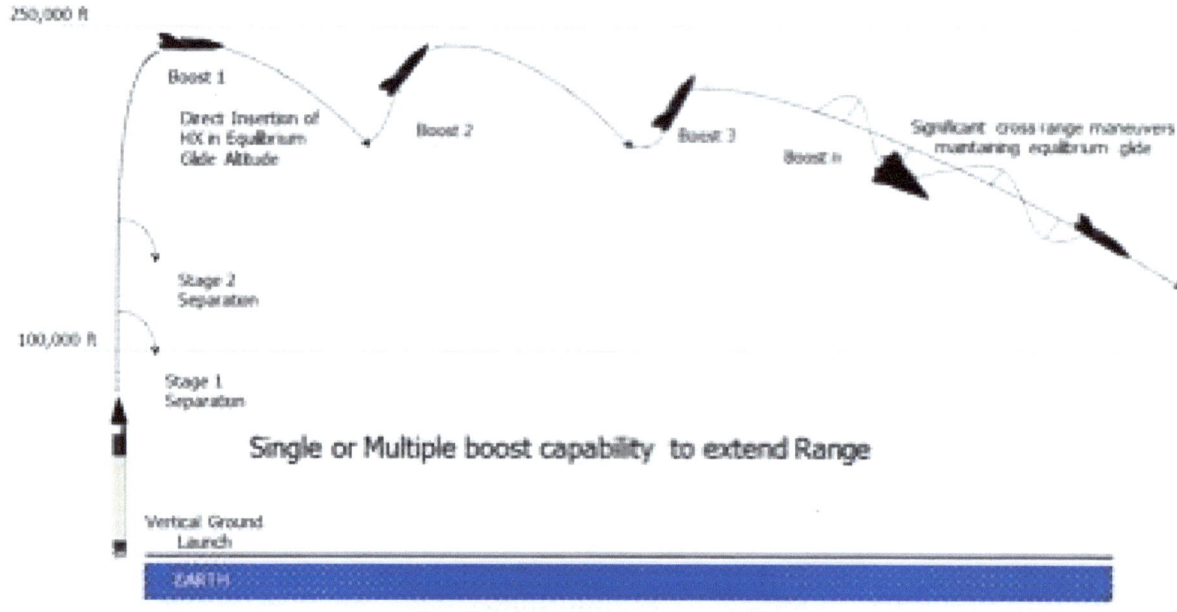

Hypersonic glide vehicle's unpredictable (controlled) flight profile

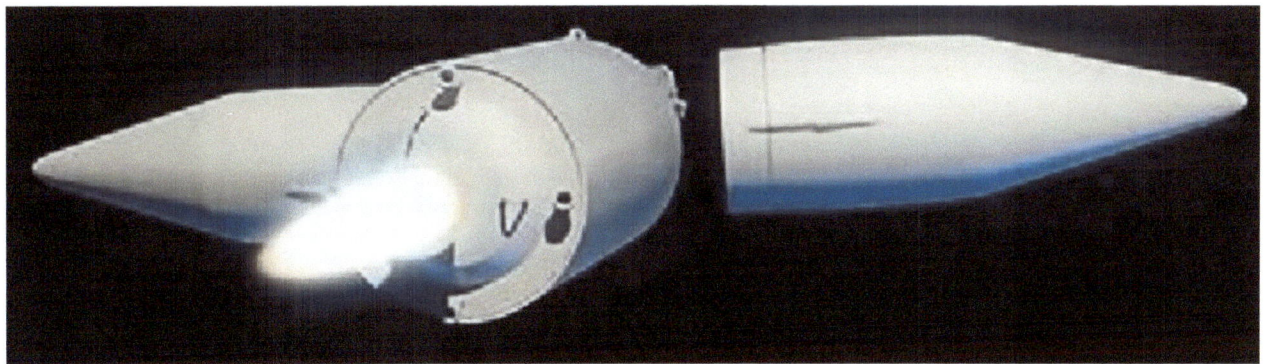

Avangard post-boost vehicle engine

SS-18 missile on tractor trailer being raised to load in silo

DARPA Artist concept (NOTE: No fins shown)

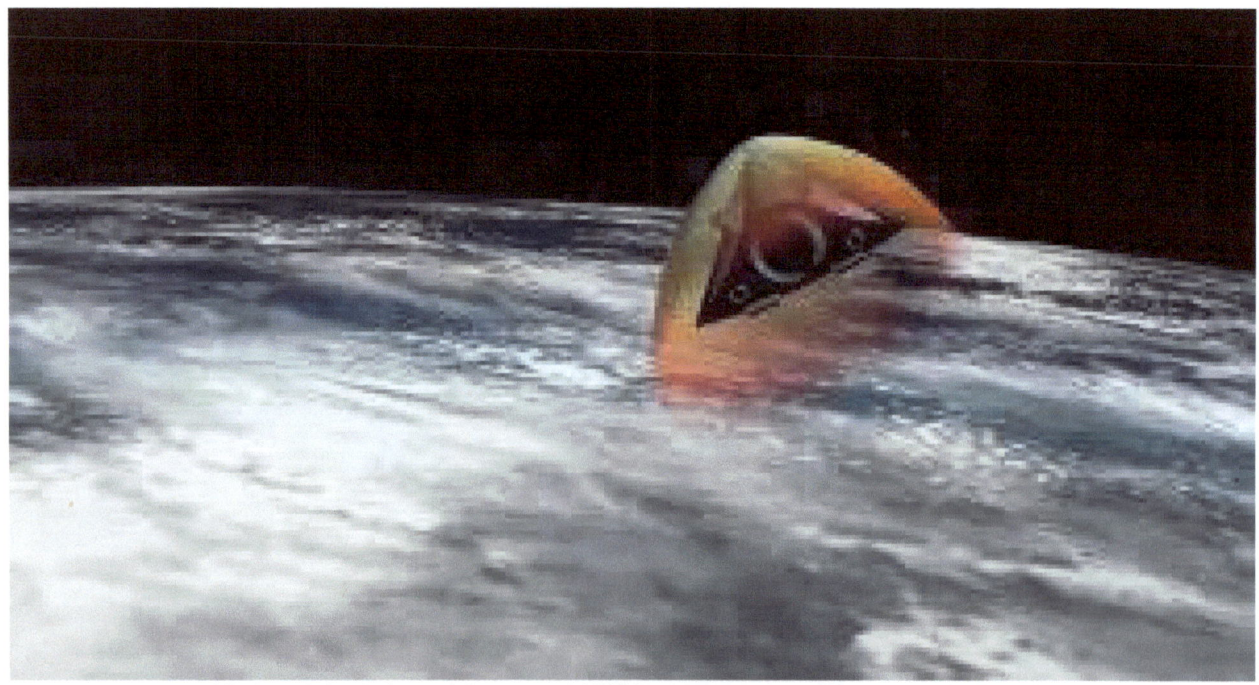
Notional depiction of Avangard HGV maneuvering. Photo credit: Ruptly

SS-N-33 Zirkon (3M-22 Hypersonic Missile)

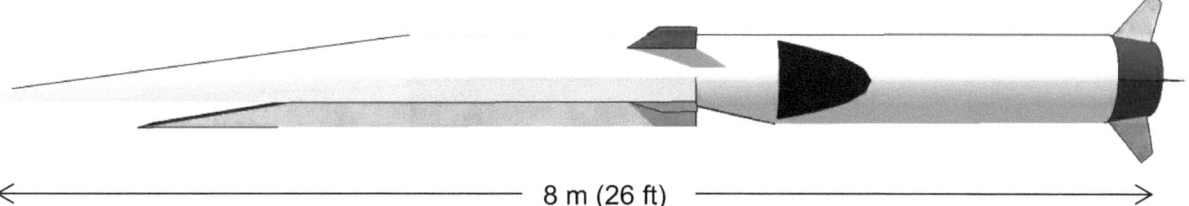

←──────── 8 m (26 ft) ────────→

The SS-N-33 (3M-22 Zircon missile; Russian: "Tsirkon") is an Intermediate Range Hypersonic Missile powered by a scramjet. A successful Zirkon test launch was held from a coastal platform in December 2018. In a test launch, the missile was clocked at a speed of Mach 8. The technical and tactical characteristics have not been disclosed and there are no known photographs of the missile (above is an artist rendering of what it is believed to look like). The missile is said to have a range of over 500 km. Over 500 km can mean 800 and 1200. Russia claims the missile can develop a speed of Mach 9 and fly at an altitude of 30-40 km where the range and speed increase as air resistance is lower. Estimated payload of 300-400 kg and a missile length at 8-10 meters (26 – 32 ft). Zirkon is to be fired from universal vertical launchers 3S-14 on warships and submarines, and from Bastion mobile coastal missile launchers.

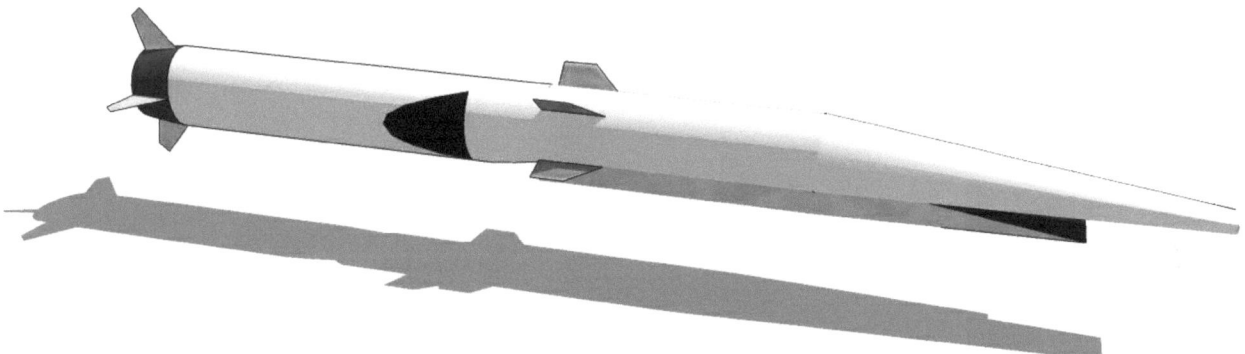

The K-300P Bastion-P (NATO reporting name SS-C-5 Stooge) missile launcher deserves some attention. A typical battery is composed of 1-2 command and control vehicles based on the Kamaz 43101 6×6 truck, one support vehicle, <u>four (4) launcher vehicles</u> based on the MZKT-7930 Astrolog 8×8 chassis (Russian: M3KT-7930) each operated by a 3-man crew and holding two missiles, and four loader vehicles; launcher vehicles can be located up to 25 km (16 mi) away from the C2 vehicles. Missiles can be readied for firing within five minutes at 2-5 second intervals. A battery can remain on active standby 3–5 days, or 30 days with a support vehicle.

(The Latest Hypersonic Rocket "Tsirkon" M-22)

НОВЕЙШАЯ ГИПЕРЗВУКОВАЯ РАКЕТА «ЦИРКОН» М-22

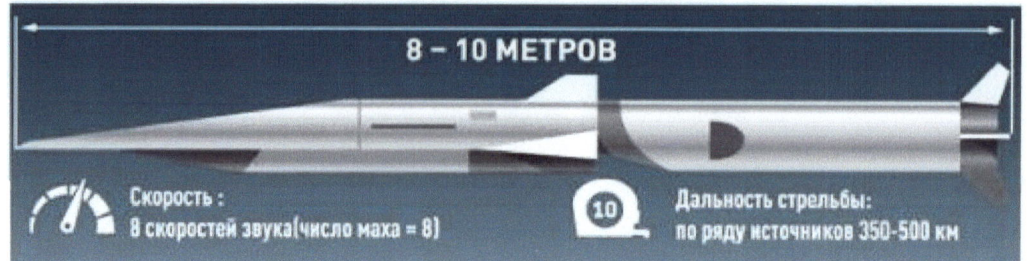

(Speed: Mach 8)   (Range: 350-500 km)

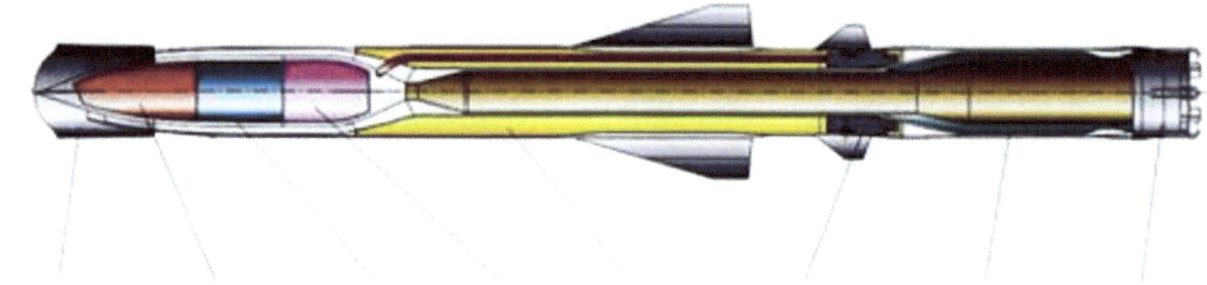

| Носовой обтекатель (nose cone) | Головка самонаведения (warhead) | Бортовая аппаратура системы управления (control system on-board equipment) | Боевой отсек (combat compartment) | Топливо маршевого двигателя (main engine fuel tank) | Рулевой привод (missile guidance gear) | Маршевый двигатель (main engine) | Стартово-разгонная ступень (launch step) |

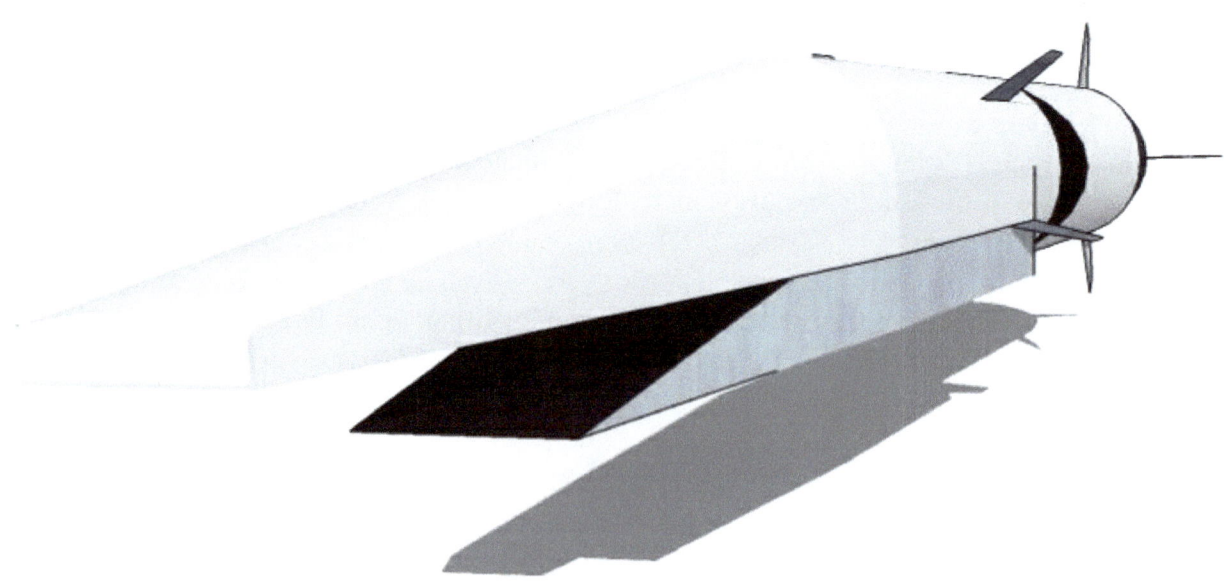

MZKT-7930 Bastion-P Transport Erector Launcher

SA-N-9 Gauntlet (Kh-47M2 "Kinzhal" ALBM)

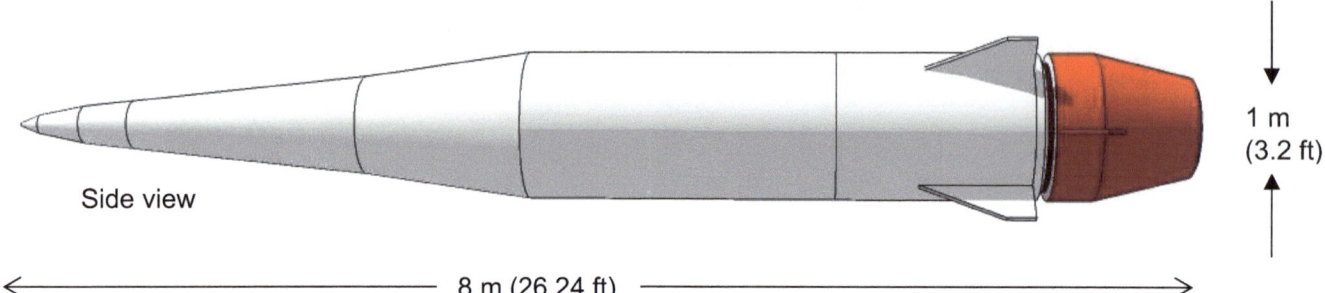

The SA-N-9 Gauntlet (Kh-47M2 Kinzhal Russian meaning "Dagger", NATO designation "Gauntlet") is a nuclear-capable air-launched ballistic missile (ALBM) with a solid-propellant. It has a claimed speed of Mach 10, a range of 3,000 km (Tu-22M3), and an ability to perform evasive maneuvers at every stage of its flight. The missile has a flight ceiling of 20 km (65,617 ft) and is controlled by an inertial guidance system with adjustments from GLONASS, remote control and an optical homing system for an accuracy of 1 m. It can carry both nuclear (100-500 Kt) or conventional 500 kg warheads and can be launched from Tu-22M3 bombers (4 missiles), MiG-31K or Su-57 aircraft. It has been deployed at airbases in Russia's Southern Military District.

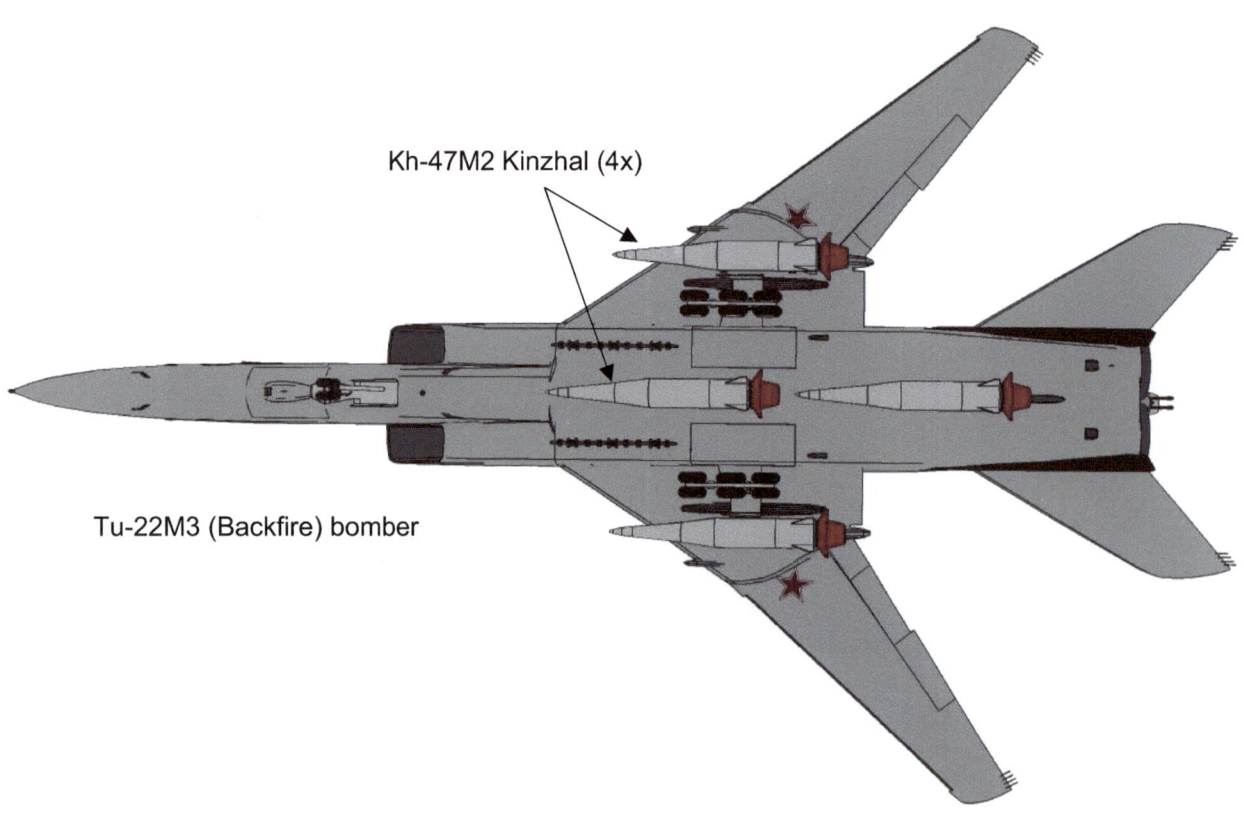

Mig-31 with Kh-47M2 "Kinzhal" air-launched ballistic missile

According to a TASS report, a MiG-31K interceptor took off from the Olenegorsk airfield in the northern Murmansk region and fired a Kinzhal missile against a ground target at the Pemboi training ground in Russia's Arctic Komi region.

LAND-BASED BALLISTIC MISSILES

Iskander-M Ballistic Missile

Test launch of the Topol-M intercontinental ballistic missile from Plesetsk base in the Arkhangelsk region.

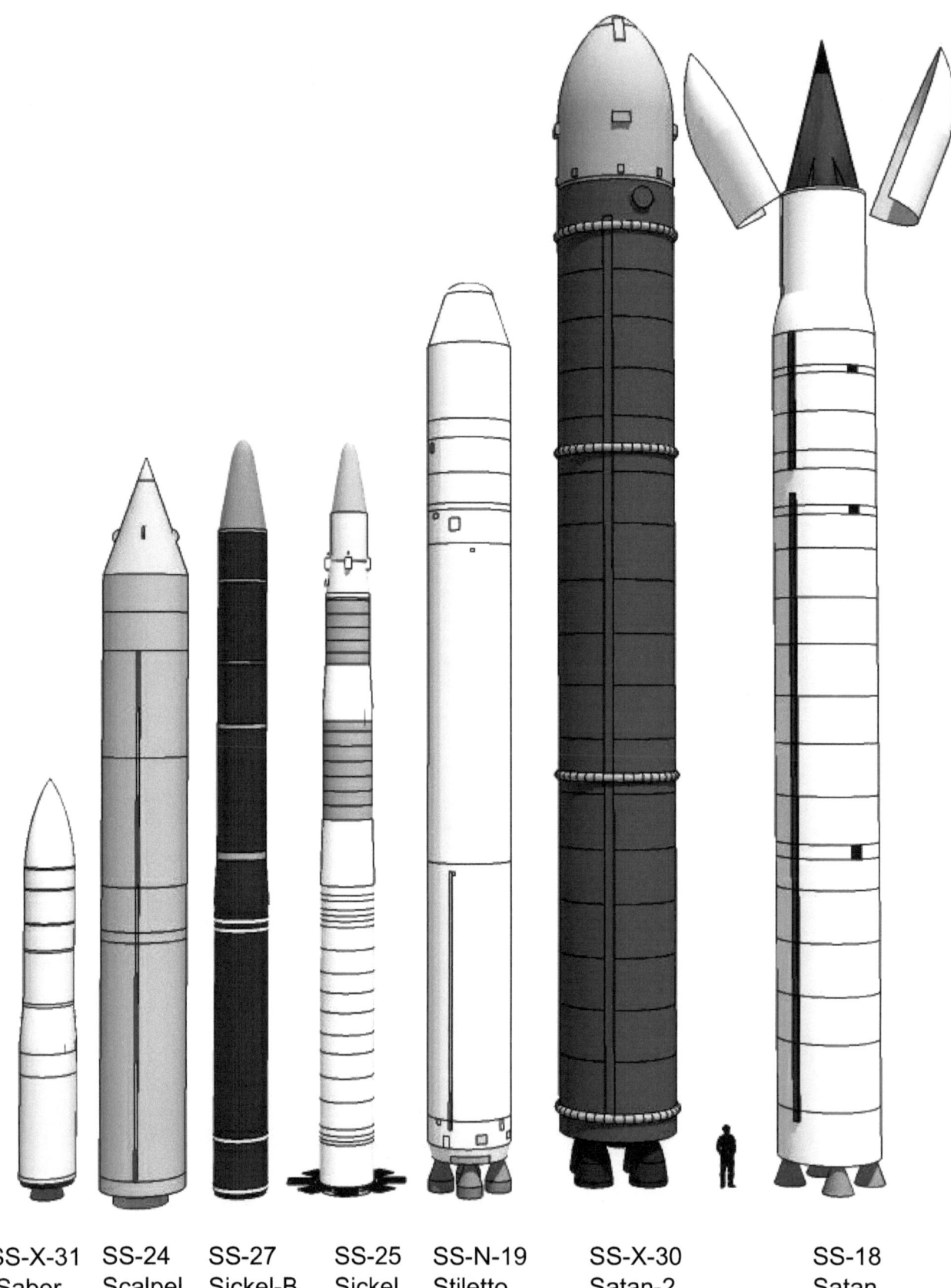

SS-18 Mod 5/6 Satan (R-36M2 "Voevoda" ICBM)

46 Launchers – 460 nuclear warheads

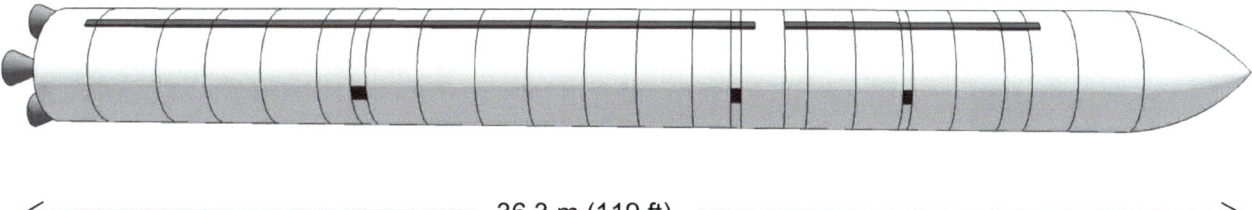

←—————— 36.3 m (119 ft) ——————→

The R-36M2 (SS-18 Mod 5; NATO reporting name: SS-18 Satan; industrial index number 15A18; Treaty designation: RS-20B) is a silo-based intercontinental ballistic missile with a two-stage rocket powered by a liquid bipropellant, using UDMH as fuel and nitrogen tetroxide as an oxidizer. The Mod 6 carries 10 MIRVs, each having a higher yield than the Mod 4 warheads (approximately 750 kt to 1 Mt). The engine of the second stage is completely built into the fuel tank (previously seen only on SLBMs) and the design of the transport-launching canister was altered. The missile is stored in a container, inserted in the silo. The container doubles as a mortar barrel - it has a drum-like "piston" beneath the filled with a gas charge that pushes the missile from the container. The empty piston is then pushed sideways so a new container with missile could be inserted in the intact silo. The maximum number deployed is 104. The R36M2 has a length of 36.3 m (119 ft) and a diameter of 3.05 m (10 ft). It has a range of 16,000 km and a CEP of 500 m. Russia intends to replace the R-36M with the SS-X-30 (RS-28 Sarmat).

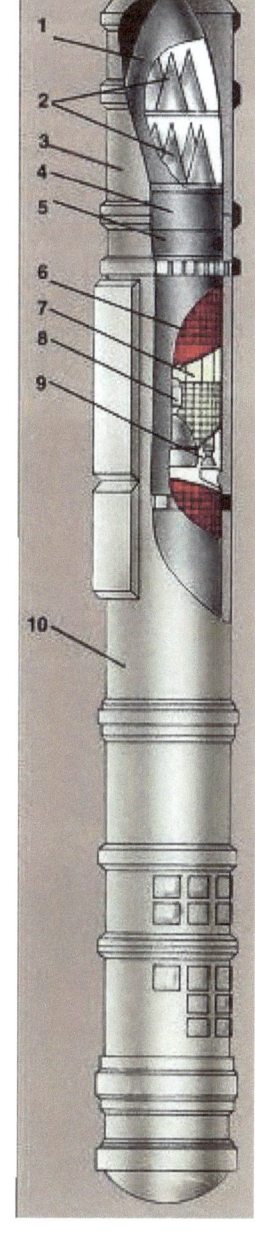

1 – nose cone
2 – reentry vehicles
3 – Transport-launch canister
4 – warhead instruments and equipment bay
5 – adapter section
6 – second-stage oxidizer tank
7 – second-stage fuel tank
8 – second-stage liquid propellant sustainer
9 – second-stage control motor gas duct
10 – transport-launch canister body

SS-19 Mod 4 Stiletto (UR-100N ICBM)

20 Launchers – 120 nuclear warheads

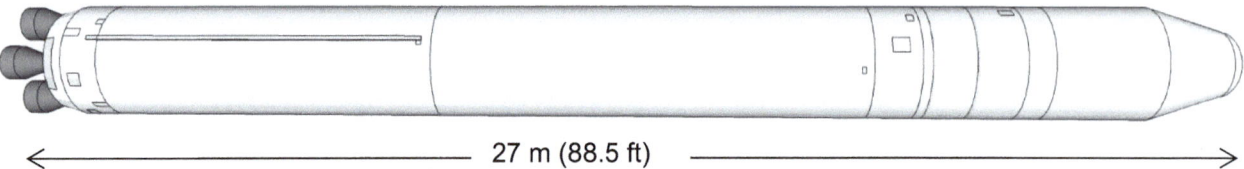

The UR-100N (MR-UR-100 "Sotka"; NATO reporting name: SS-19 "Stiletto"; industry designation 15A30) is a fourth-generation silo-launched liquid-propellant (asymmetrical dimethylhidrazine and nitrogen tetraoxide) intercontinental ballistic missile in service with the Russian Strategic Missile Troops (13th Red Banner Rocket Division at Yasny, Orenburg Oblast). The SS-19 Mod 3 holds up to 6 MIRVs. The HGV-carrying version of the missile has the designation **SS-19 Mod 4**. The missile has a length of 27 m (88.5 ft) and a diameter of 2.5 m (8.2 ft). It has a range of 10,000 km and a CEP of 350-550 m. The missile has a preparation time to start of 25 minutes, a storage period of 22 years, and carries 6 MIRVs, each yielding between 550 – 750 kilotons. The first missile regiment armed with the Avangard HGV entered service on 27 December 2019. The missile is also known as the RS-18A. As of 2016, twenty (20) SS-19 launchers were in service.

SS-19 before it is dismantled in Ukraine in 1999. Photo credit: Taiwan News

SS-21 Scarab (OTR-21 "Tochka" SRBM)

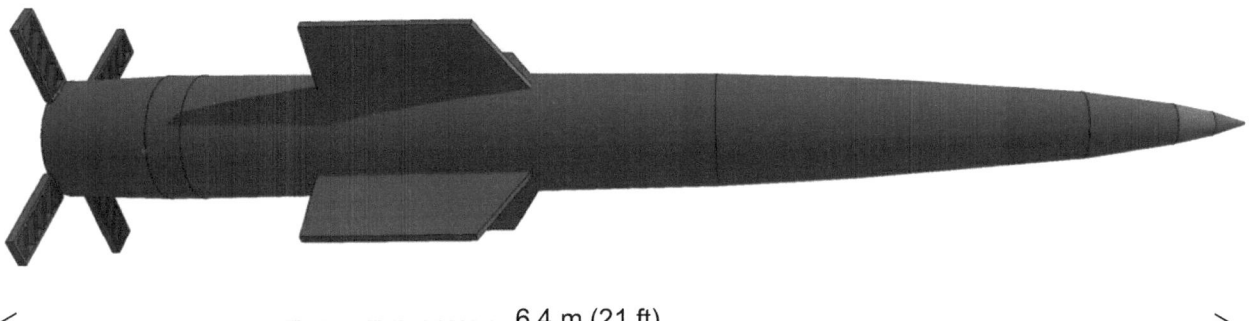

←————————— 6.4 m (21 ft) —————————→

The SS-21 Scarab (OTR-21 "Tochka", Russian: оперативно-тактический ракетный комплекс, English: Tactical Operational Missile Complex Tochka, NATO designation: SS-21 "Scarab") is a tactical ballistic missile with a solid-propellent. It has a length of 6.4 m (21 ft) and a diameter of 0.65 m (2 ft 2 in). It uses an inertial guidance system and the missile range varies: Scarab A, 70 km (43 mi) with CEP of 150 m (490 ft), Scarab B, 120 km (75 mi) with CEP of 95 m (312 ft); Scarab C, 185 km (115 mi) with CEP of 70 m (229 ft). Its GRAU designation is 9K79. It is transported in a 9P129 amphibious vehicle and raised prior to launch.

9P129 6×6 transport erector launcher with amphibious capability.
Max road speed is 60 km/h (37 mph), and 8 km/h (5.0 mph) in water.

SS-24 Scalpel (RT-23 "Molodets" ICBM)

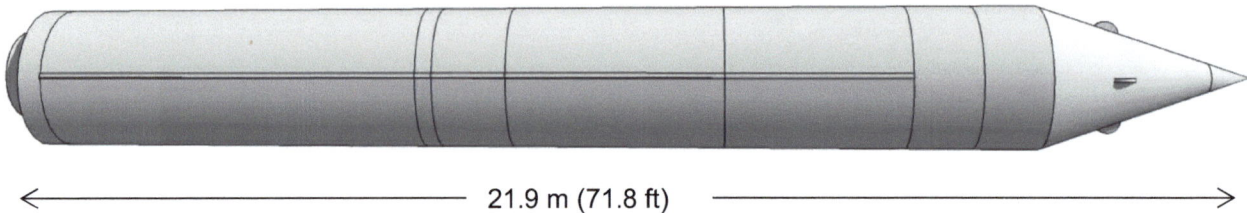

←——————— 21.9 m (71.8 ft) ———————→

The SS-24 Scalpel (RT-23 "Molodets", Russian: РТ-23 УТТХ "Мо́лодец", lit. "brave man" or "fine fellow"; NATO reporting name: SS-24 "Scalpel") was a cold-launched, three-stage, solid-fueled intercontinental ballistic missile. The sustainer-stage propellant was a solid-propellant mixture, and the propellants in combat stage were liquid, high-boiling hypergolic: UDMH fuel and NTO oxidizer. The missile has a length of 21.9 m (71.8 ft) and a diameter of 2.4 m (7.8 ft). It came in silo- and rail-based variants, and was armed with 10 MIRV warheads (GRAU index: 15Ф444) of 550 kT yield. A battery consisted of three 3-car launch modules with missiles; 7-car command module; tank car with fuels and lubricants stock; 2 diesel locomotives. The missile included a hinged control nozzle used as a thrust-vectoring element. All missiles were decommissioned by 2005 in accordance with the START II. It has been suggested that these rail-mobile land-based missiles, which have been parked in their garrisons, may be placed back on patrol in response to American missile defense and associated arms control initiatives.

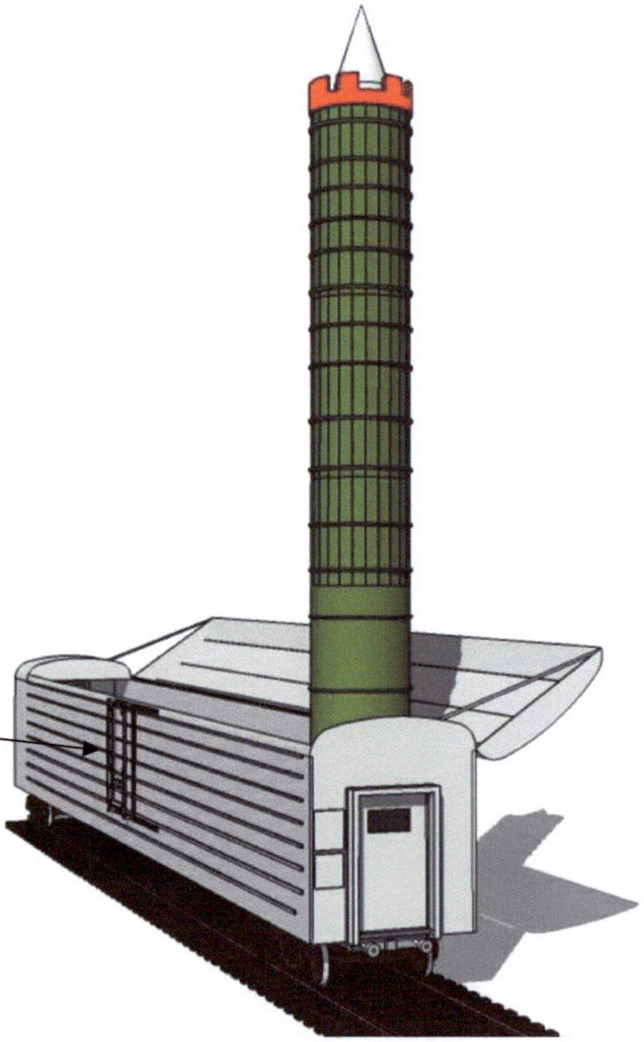

False side doors disguised to resemble refrigerated wagons

SS-25 Sickle (RT-2PM "Topol" ICBM)

72 Launchers – 72 nuclear warheads

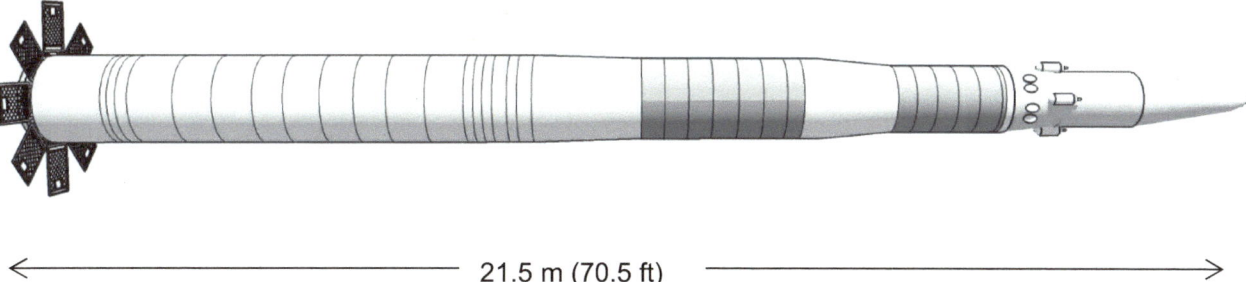

21.5 m (70.5 ft)

The SS-25 Sickle (RT-2PM "Topol", Russian: РТ-2ПМ Тополь; meaning "Poplar"; NATO reporting name SS-25 "Sickle"; GRAU designation: 15Ж58 ("15Zh58"); START I designation: RS-12M Topol is a road-mobile intercontinental ballistic missile with a three stage solid propellant in service with Russia's Strategic Missile Troops. It has a length of 21.5 m (70.5 ft) and a diameter of 1.7 m (5.5 ft). It carries a single warhead with a yield of 800 kt and has a CEP of 200 m. All three stages are made of composite materials. During first stage operation the flight control is implemented through four aerodynamic and four jet vanes. Four similar trellised aerodynamic surfaces serve for stabilization. During the second and third stage of flight gas is injected into the diverging part of the nozzle for flight control. By the early 2020s, all SS-25 ICBMs will be replaced by versions of the SS-27 Topol-M.

Launch legs (stowed)   Automatic gyrocompass

MAZ-7912 14×12 Twelve-wheel drive transporter-erector-launcher (710 hp V-58-7 diesel engine) two single-place cabins made of fiberglass: the left one for the driver, the right one — for the crew commander. Range — 440 km.

SS-26 Stone (9K720 "Iskander" SRBM)

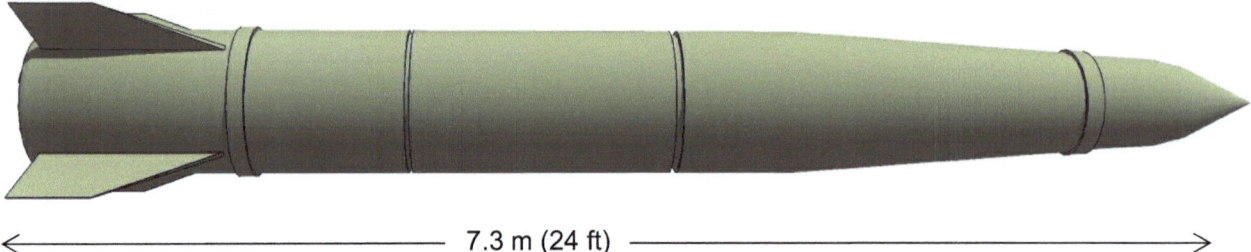

←——————— 7.3 m (24 ft) ———————→

The SS-26 Stone (9K720 "Iskander", Russian: «Искандер»; NATO reporting name SS-26 "Stone") is a mobile short-range ballistic missile system. The missile systems (Искандер-М) are to replace the obsolete OTR-21 Tochka systems, still in use by the Russian armed forces, by 2020. The missile has a length of 7.3 m (24 ft) and a diameter of 0.92 m (3 ft 0 in) and has a range of 50 km (31 mi) and 400–500 km (250–310 mi) for Iskander-M. The Iskander has several different 480–700 kg (1,060–1,540 lb) conventional warheads, including a cluster munitions warhead, a fuel-air explosive enhanced-blast warhead, a high explosive-fragmentation warhead, an earth penetrator for bunker busting and an electromagnetic pulse device for anti-radar missions. The missile can also carry nuclear warheads. The missile has a speed of 2,000 m/s (Mach 5.9). In September 2017, the KB Mashinostroyeniya (KBM) general designer Valery M. Kashin said that there were at least seven types of missiles (and "perhaps more") for Iskander, including one cruise missile.

SS-27 Mod 1 Sickle B (RT-2PM2 "Topol-M" ICBM)

Mod 1 – 78 Launchers – 72 nuclear warheads

Mod 2 – 102 Launchers – 408 nuclear warheads

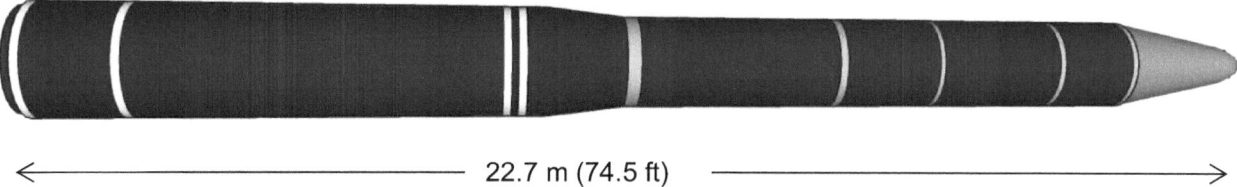

←——————— 22.7 m (74.5 ft) ———————→

The SS-27 Mod 1 Sickle B (RT-2PM2 "Topol-M", Russian: РТ-2ПМ2 "Тополь-М", NATO reporting name: SS-27 "Sickle B", other designations: SS-27 Mod 1, Treaty designation: RS-12M1, RS-12M2) is one of the most recent intercontinental ballistic missiles to be deployed and the first to be developed after the dissolution of the Soviet Union. It was developed from the RT-2PM Topol mobile intercontinental ballistic missile as a response to the American Strategic Defense Initiative. It is a solid-fuel rocket 22.7 m (74.5 ft) in length and the first stage has a body diameter of 1.9 m (6.2 ft). The body of the rocket is made by winding carbon fiber. The Topol-M carries a single warhead with a 1 Mt yield, but it is capable of carrying four to six MIRV warheads along with decoys. It has a minimum range of 2,000 km and a maximum range 10,500 km with a CEP of 200 m. It has three solid rocket stages with inertial, autonomous flight control utilizing an onboard GLONASS receiver. It was designed to counter and evade current or planned U. S. missile defense systems. It is said to be capable of making evasive maneuvers to avoid a kill by interceptors, and carries targeting countermeasures and decoys. The missile has a short engine burn time following take-off, intended to minimize detection of launch by satellite and has a much higher acceleration than other ICBMs. This enables the missile to accelerate to the speed of 7,320 m/s and to travel a flatter trajectory to distances of up to 10,000 km. The warhead is capable of changing course after separating from the launcher, making it difficult to predict a re-entry trajectory. As of March 2020, 78 RT-2PM2 Topol-M missiles are deployed with 2 rocket divisions: 60th Rocket Division at Tatishchevo Air Base (60 silo-based) and the 54th Guards Rocket Division at Teykovo (18 road-mobile).

MZKT-79221 "Universal" 16x16 wheeled transporter-erector-launcher (800 horse power V12 diesel engine). Maximum speed - 40 km/h. Radius of turn - 25 meters.

SS-29 (RS-24 "Yars" ICBM)

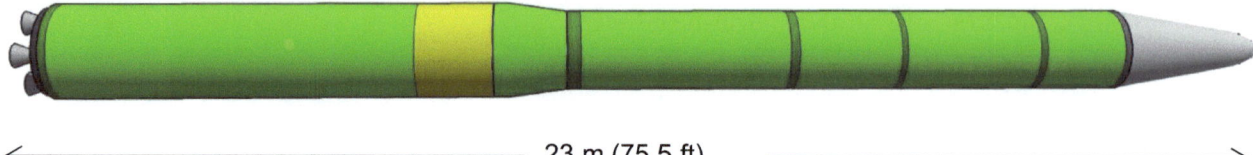

<—————————— 23 m (75.5 ft) ——————————>

The SS-29 (RS-24 "Yars", Russian: Yadernaya Raketa Sderzhivaniya, Ядерная ракета сдерживания, meaning "Nuclear Deterrence Rocket", NATO reporting name: SS-29 or SS-27 Mod 2) is a Russian MIRV-equipped, thermonuclear-armed, four-stage intercontinental ballistic missile. Russia claims the RS-24 was a completely new ICBM to justify the designation SS-29 instead of SS-27 Mod 2, to circumvent START treaty prohibition on the maximum number of warheads attributed to ICBMs. The missile can carry at least 3-4 MIRVs with 300—500 kiloton, 4-6 with 150 kiloton warheads. It was first tested on May 29, 2007 and is intended to replace the older R-36 and UR-100N. The missile uses a solid propellant (third or fourth stage can be liquid). The missile is 23 m in length (75.5 ft), 2 m in diameter (6.5 ft) and has a range of 11,000 km (6,800 mi)-12,000 km (7,500 mi). It uses an inertial guidance system with Glonass and has a CEP of 150 m. It can be launched from a silo or a road-mobile TEL MZKT-79221 and the missile can reach a speed of Mach 20. It is also known as the Topol'-MR.

Who doesn't love a parade?

SS-X-30 Satan-2 (RS-28 "Sarmat" ICBM)

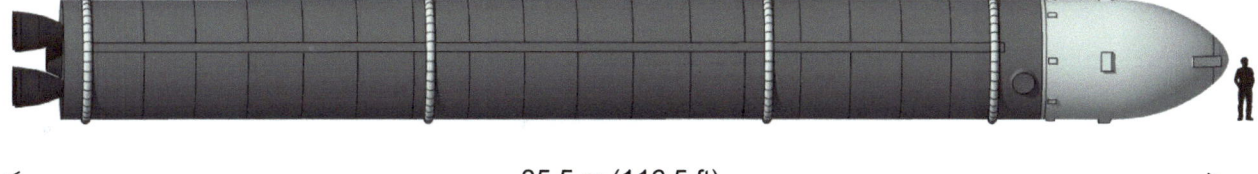

<— 35.5 m (116.5 ft) —>

The SS-X-30 Satan-2 (RS-28 "Sarmat", Russian: РС-28 "Сармат", NATO reporting name: SS-X-29 or SS-X-30 "Satan-2") is a liquid-fueled, MIRV-equipped heavy intercontinental ballistic missile under development by the Makeyev Rocket Design Bureau since 2009. It is intended to replace the R-36M ICBM (SS-18 "Satan"). The missile has a length of 35.5 m (116.5 ft) and a diameter of 3.5 m (11.5 ft. It uses inertial guidance, GLONASS, Astro-inertial system and has a CEP of 10m. It carries more nuclear warheads than its predecessor which had up to 10 heavy or 15 light MIRV warheads and 40 penetration aids. Sarmat has a short boost phase, which shortens the interval when it can be tracked by satellites with infrared sensors. Russia claims the Sarmat's range of 18,000 km (11,000 mi) allows it to fly over both the North and the South poles to reach any target. This is important because it is speculated that the Sarmat could fly a trajectory over the South Pole, completely immune to any current missile defense system, and that it has the Fractional Orbital Bombardment (FOBS) capability. FOBS was developed by the Soviets in the 1960s to use space to deliver nuclear weapons into low Earth orbit before bringing them down on their targets. It requires a launch vehicle powerful enough to be capable of putting the weapon into orbit. However, the orbit is only a fraction of a full orbit, not sustained, and so there will be much less need to control a precise orbit. The maximum altitude would be around 150km.

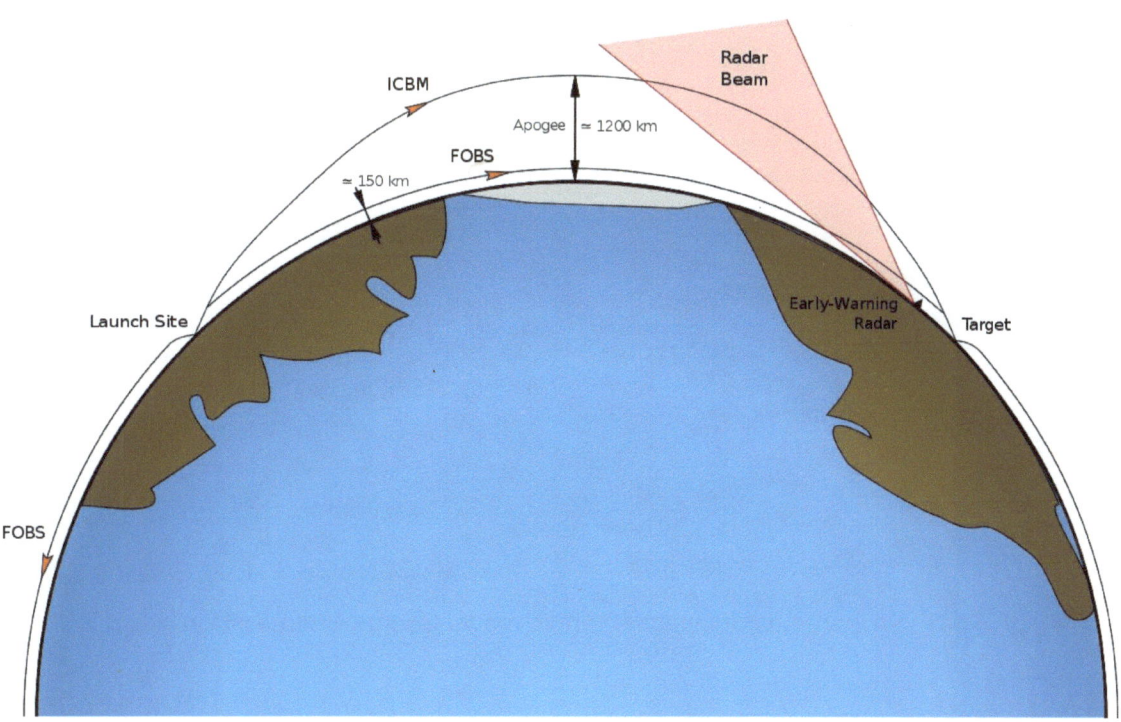

SS-X-31 Saber (RS-26 "Rubezh" ICBM)

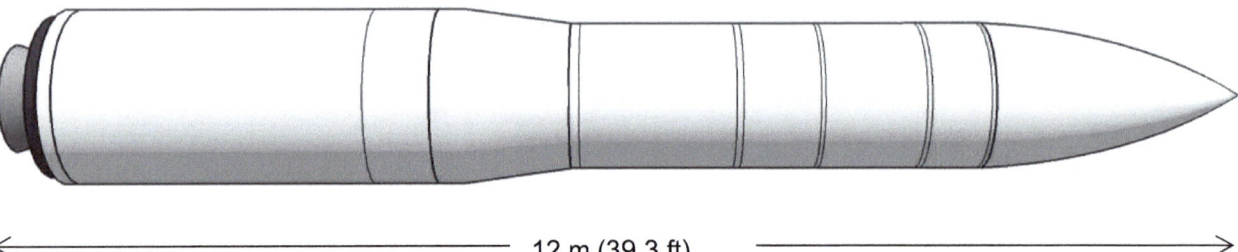

<— 12 m (39.3 ft) —>

The SS-X-31 Saber (RS-26 "Rubezh", Russian: РС-26 Рубеж; NATO reporting name: SS-X-31 or SS-X-29B "Saber") is a solid-fueled, three-stage intercontinental ballistic missile, equipped with 4 thermonuclear MIRV or MaRV (Maneuverable Reentry Vehicle) payloads (150-300 Kt each). The missile is also known under the name of its R&D program Avangard (Авангард) intended to be capable of carrying the Avangard hypersonic glide vehicle.[1] The RS-26 is a shorter version of the RS-24 (one fewer stage). The missile length (without Avangard) is 12 m (39.3 ft) and 1.8 m (5.9 ft) in diameter. A modified nose cone and fairing would be required if it is to carry an Avangard and the length would probably increase to at least 17 m (55.7 ft). The missile has a demonstrated range of 5,800 km (3,600 mi) with a light or no payload. This is above the agreed 5,500 km limit of the INF treaty. In 2018, it was reported that development of the RS-26 was *frozen until at least 2027*, and funding diverted toward continued development of the Avangard hypersonic glide vehicle.

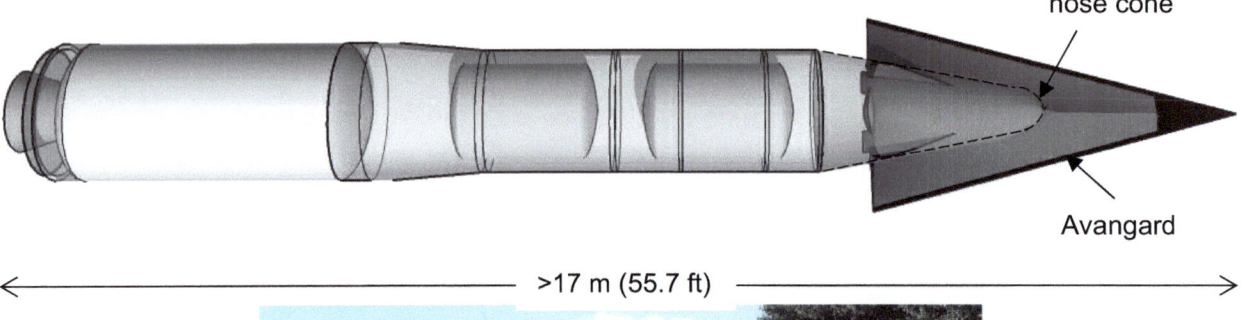

<— >17 m (55.7 ft) —>

MAZ-547 transport-erector launcher (photo credit: George Chernilevsky)

---

[1] I don't see how it would be possible to install Avangard on this missile as it clearly would not fit.

SS-X-32Zh Scalpel B (RS-27 "Barguzin" ICBM)

The SS-X-32Zh Scalpel B (RS-27 "Barguzin" BZhRK, Russian: БЖРК; NATO reporting name: SS-X-32Zh "Scalpel B") Project is a rail-mobile intercontinental ballistic missile that was under development for the Russian RVSN, as a replacement of the previous railway missile train Molodets BZhRK SS-24 Scalpel. BZhRK stands for railway strategic missile train. The new Ghost Train would use much lighter RS-24 Yars missile that weighed only fifty-four tons, and carried only four nuclear warheads. The new system therefore could use standard train wagons with regular wheels and mount six ballistic missiles instead of three. By comparison, before launch, the RT-23 Molodets would have to stop at one of two hundred specially-constructed firing points along the rail system, shove aside overhanging power cables with a special device, and pop open the roofs of the missile trains. To avoid frying the train with rocket exhaust, the Molodets used a 'cold-launch' system: a powder charge propelled the missile twenty meters into the air, then a thruster tilted the missile away from the train before its rocket boosters finally ignited. The missile was expected to enter testing in 2019 and enter service in 2020, but the Barguzin was ultimately deemed too dated and expensive given a constricted defense budget.

False side doors disguised to resemble refrigerated wagons.

Photo of the SS-24 Scalpel (RT-23 "Molodets" ICBM), not the SS-X-32Zh Scalpel B.

SUBMARINE-LAUNCHED BALLISTIC MISSILES

Bulava launched from submarine Yuri Dolgoruky on 28 October 2011

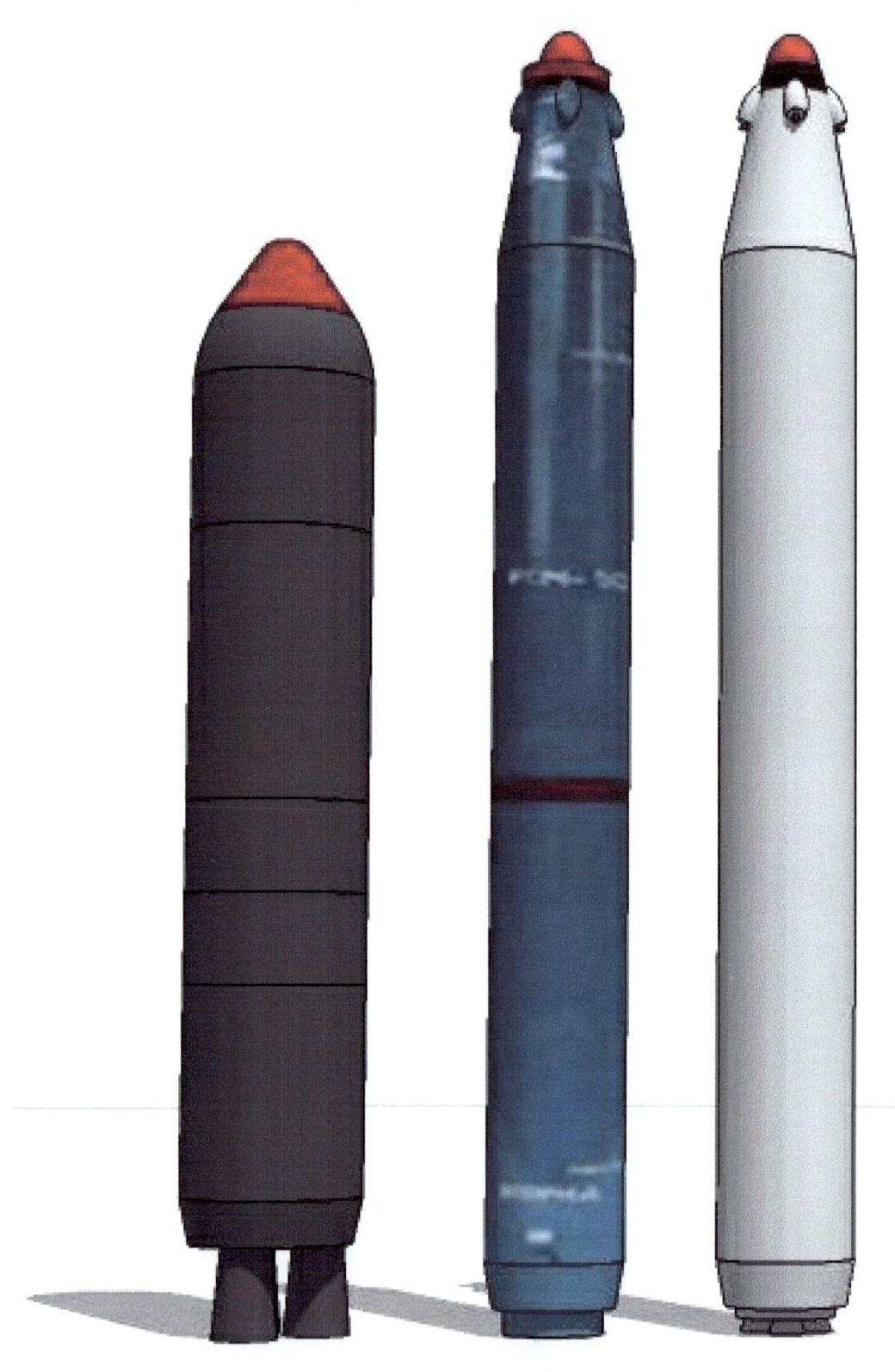

SS-NX-30
Bulava

SS-N-18
Stingray

SS-N-23
Skiff

SS-N-18 Stingray (R-29R "Vysota" SLBM)

2/32 Launchers – 96 nuclear warheads

←——————————— 14.4 m (47.2 ft) ———————————→

The SS-N-18 (R-29R "Vysota" Р-29 Высота, NATO reporting name: SS-N-18 "Stingray", SALT designation: RSM-50) is a submarine-launched, two-stage intercontinental-range, liquid-propellant ballistic missile. It has a range of 6,500 km (4,038 miles) and likely derives from the SS-N-8 (R-29) missile. The R-29R missile has a length of 14.4 m (47.2 ft) and a diameter of 1.8 m (5.9 ft) and was designed for the Delta III ballistic missile submarine (SSBN), or Kalmar class. Each Kalmar class submarine carries sixteen SS-N-18 missiles.

K-18 "Karelia" Delta III SSBN (photo credit: US Navy)

SS-N-23 Skiff (R-29RMU2 "Layner" SLBM)

6/96 Launchers – 384 nuclear warheads

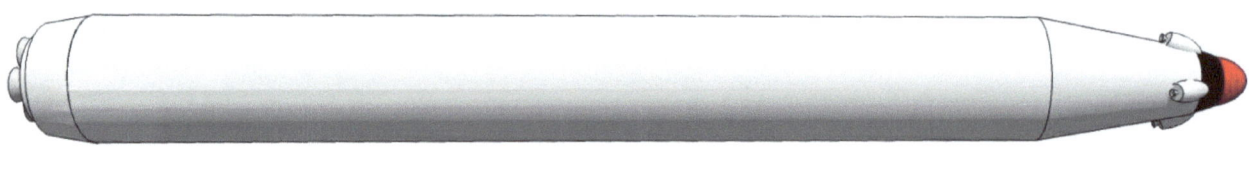

14.8 m (48.55 ft)

The SS-N-23 Skiff (R-29RMU2.1 "Layner" Russian: Р-29РМУ2 Лайнер meaning Liner, NATO designation "Skiff", GRAU index 3M27) is a liquid-fueled three-stage Submarine-Launched Ballistic Missile (SLBM) and the newest member of the R-29 missile family, developed by the Makeyev Rocket Design Bureau and produced by the Krasnoyarsk Machine-Building Plant. The missile has a length of 15 m (49.2 ft), a diameter of 1.9 m (6.2 ft) and has a range of 8,300-12,000 km (5,157-7,456 mi.). Derived from the R-29RMU "Sineva" SLBM, the Layner can carry twelve nuclear warheads, three times as many as Sineva. The warheads can be of various yields with fewer warheads. The missile can carry 12 low-yield warheads without penetration aids, 10 low-yield warheads with penetration aids, 8 low-yield warheads with enhanced penetration aids, or 4 medium-yield warheads with penetration aids. While it shares flight characteristics with the Sineva, the Layner is equipped with improved systems to overcome anti-ballistic missile shields. It entered service with the Russian Delta IV class submarines in 2014.

SS-NX-30 (RSM-56 "Bulava" SLBM)
3/48 Launchers – 288 nuclear warheads

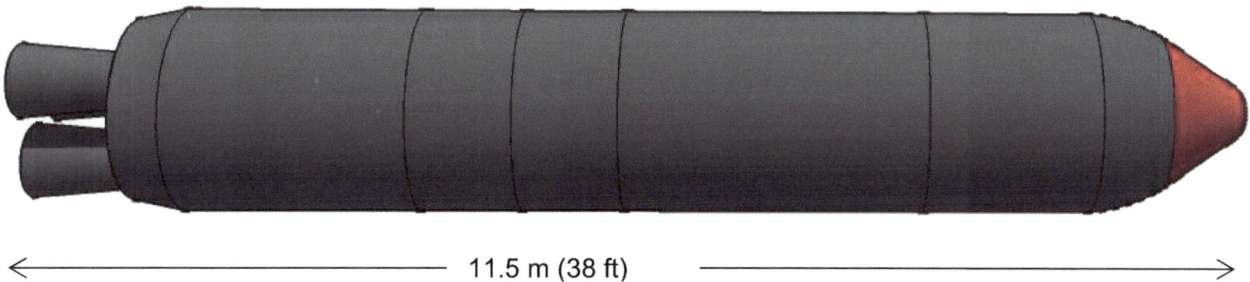

←———————— 11.5 m (38 ft) ————————→

The SS-NX-30 (RSM-56 "Bulava", Russian: Булава, meaning "Mace", NATO reporting name: SS-NX-30 or SS-N-32, GRAU index 3M30, 3K30) is a submarine-launched ballistic missile (SLBM) deployed in 2013 on the Borei class of ballistic missile nuclear submarines. The missile has a length without warhead of 11.5 m (38 ft) and a diameter of 2 m (6 ft 7 in). It has a range of 8,000-8,300 km (4,970-5,157 mi.) with a CEP of 350 m. The missile uses inertial guidance, possibly with Astro-inertial guidance and/or GLONASS. It is a cornerstone of Russia's nuclear triad, and is the most expensive weapons project in the country. The Project 955/955A Borei-class submarines carry 16 missiles per vessel. Development and deployment of the Bulava missile within the Russian Navy is not affected by the enforcement of the new START treaty. Bulava is both lighter and more sophisticated than the Topol-M. The missile has three stages; the first and second stages use solid fuel propellant, while the third stage uses a liquid fuel to allow high maneuverability during warhead separation. The missile can be launched from an inclined position, allowing a submarine to fire them while moving. It has a low flight trajectory, and due to this could be classified as a quasi-ballistic missile. It possesses advance missile defense evasion capabilities making it resistant to any missile-defense systems. Bulava can be loaded on TEL road mobile launchers, on railway BZhRK trains and other various launchers. The missile's flight test program was experienced problems due to substandard components. Russian military expert Pavel Felgenhauer said that of the Bulava's first 12 test launches, only one was entirely successful.

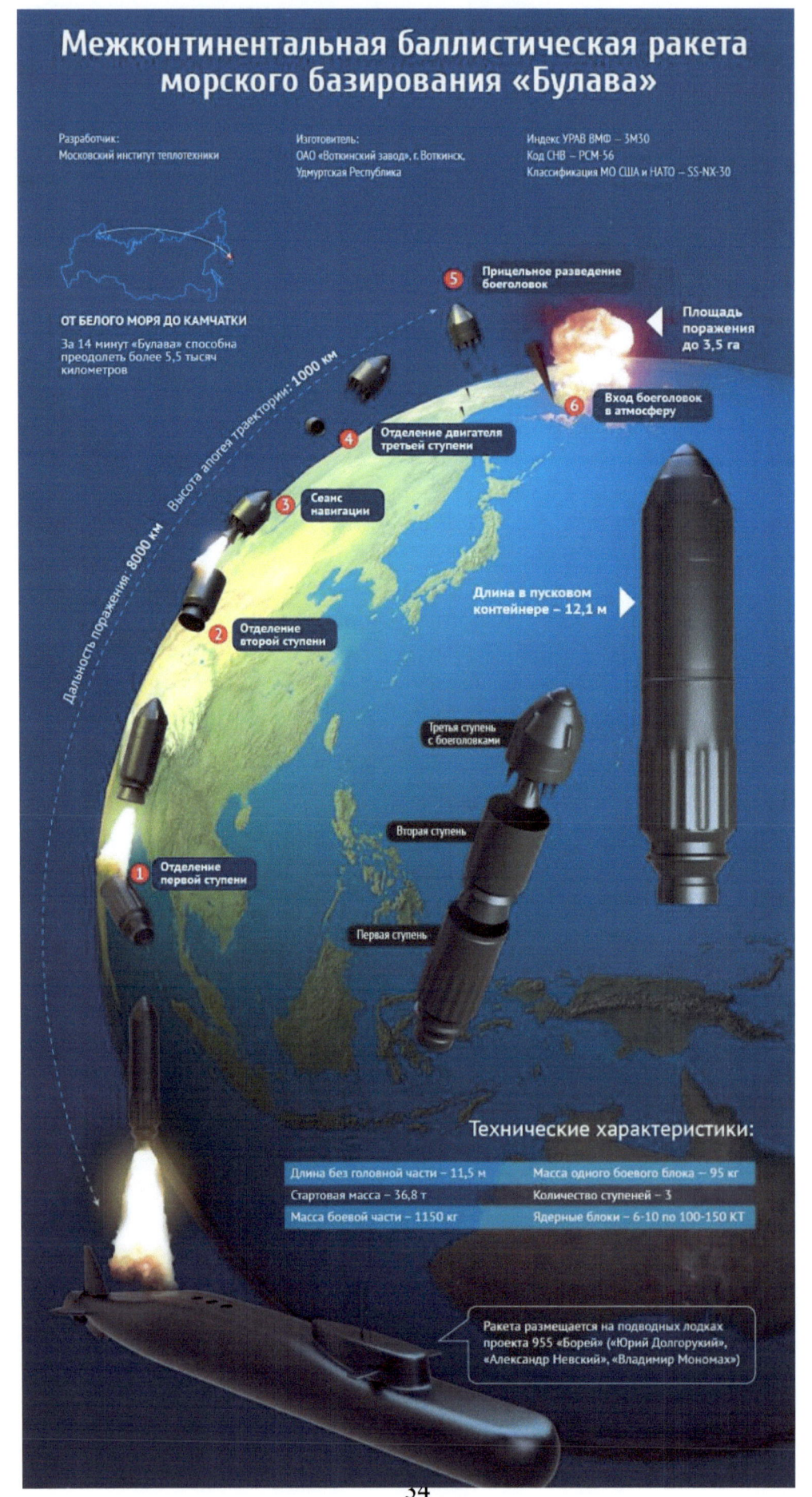

CRUISE MISSILES

Российский истребитель пятого поколения Т-50 и ракета Х-35УЭ

Истребитель Т-50

Экипаж	1 пилот
Длина	19,7 м
Размах крыла	14 м
Макс. взлетная масса	37 000 кг
Предельная скорость	2600 км/ч
Боевая нагрузка	10 000 кг

Ракета Х-35УЭ

Предназначена для поражения боевых надводных кораблей и наземных целей

Дальность	260 км
Средняя скорость полета	300 м/с
Диапазон высот пуска	0,2–10 км
Макс. дальность захвата цели	50 км
Масса боевой части	145 кг

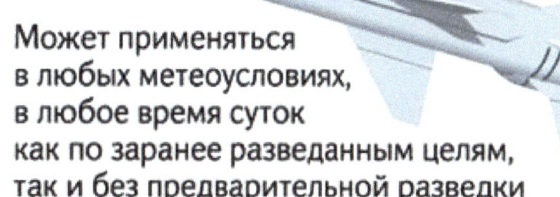

Может применяться в любых метеоусловиях, в любое время суток как по заранее разведанным целям, так и без предварительной разведки

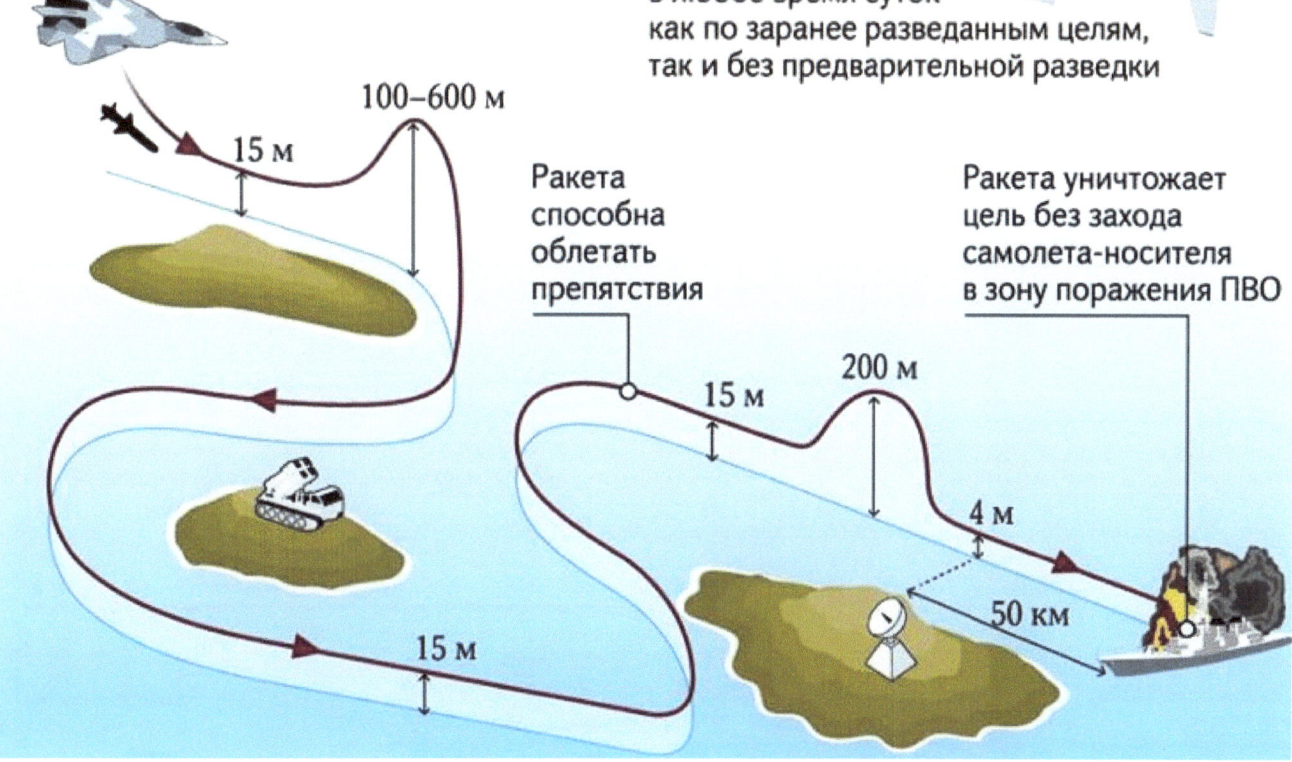

Ракета способна облетать препятствия

Ракета уничтожает цель без захода самолета-носителя в зону поражения ПВО

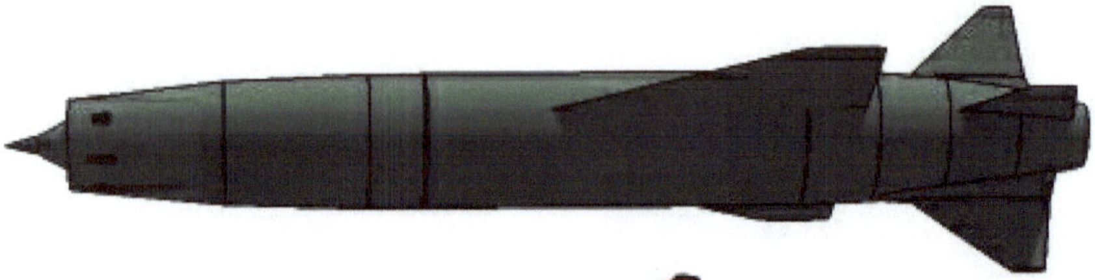

SS-N-19 Shipwreck

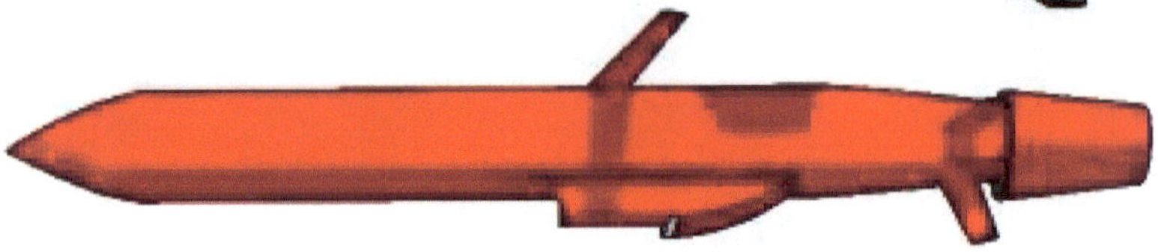

SSC-X-9 Skyfall

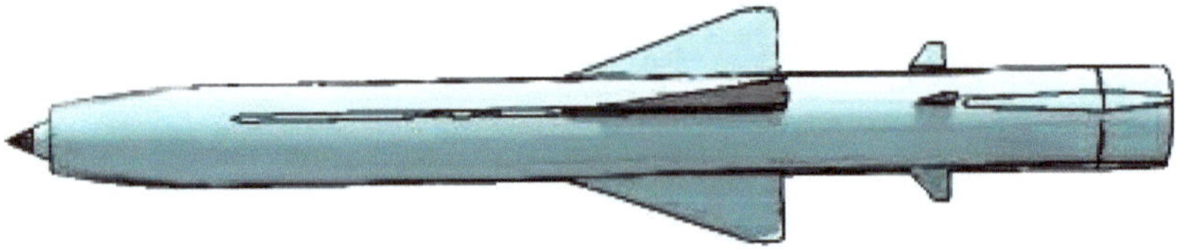

SS-N-26 Strobile

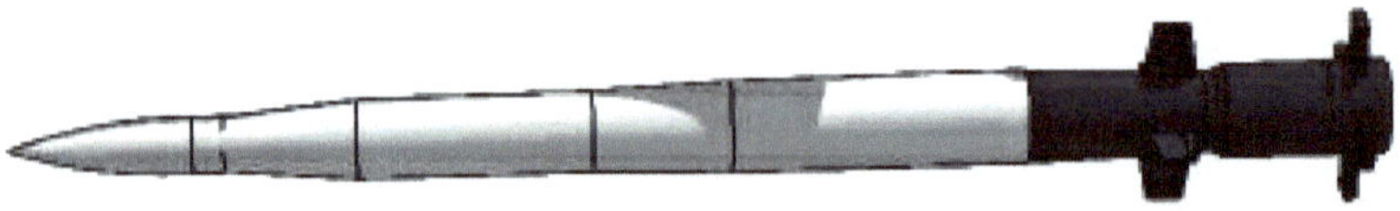

SS-N-27 Sizzler

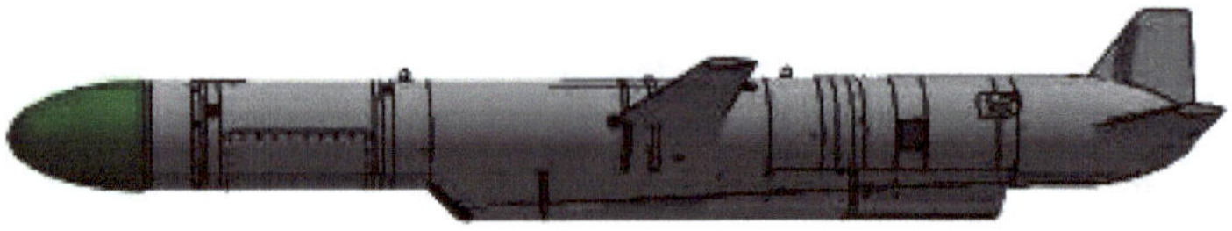

SS-N-21 Sampson

AS-23A Kodiak

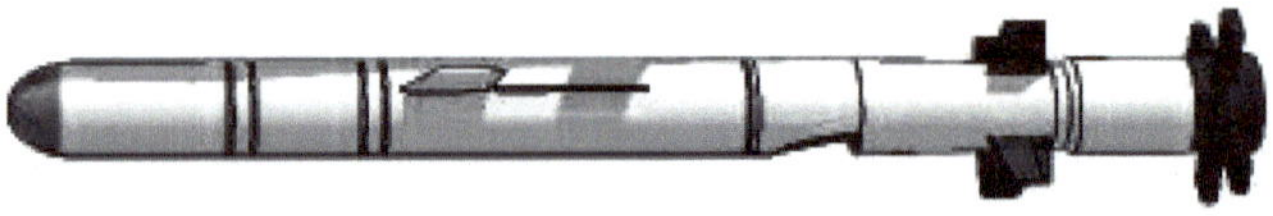

SS-N-30A

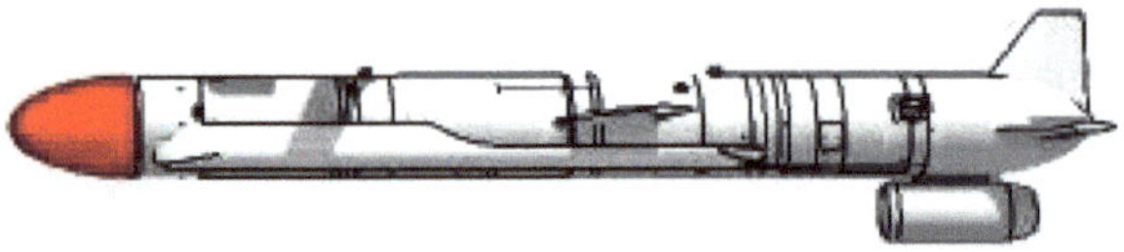

AS-15 Kent

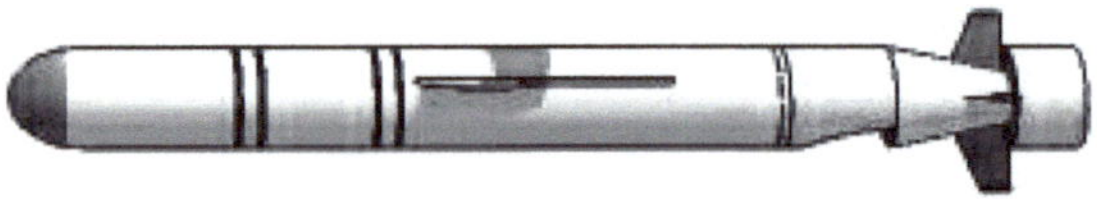

SSC-8 Screwdriver

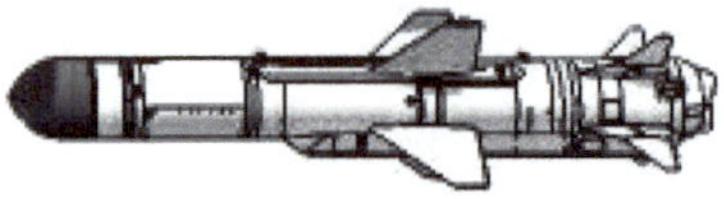

SS-N-25 Switchblade

SSC-8 Screwdriver (9M729 GLCM)

16 Launchers – 16 nuclear warheads

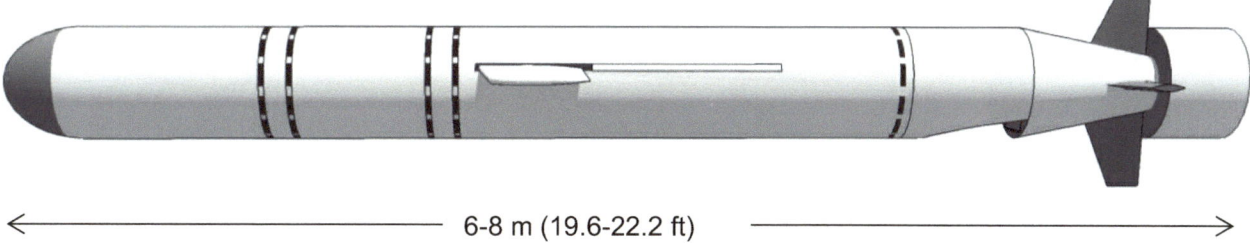

← 6-8 m (19.6-22.2 ft) →

The SSC-8 is a ground-launched cruise missile approximately 6-8 m (19.6-22.2 ft) in length and 0.533 m (1.75 ft) in diameter. Although Russia claims that it has a range of under 482 km (300 mi.), it has reportedly been tested at various ranges and is believed to have a maximum range of 2,500 km. The missile employs a guidance system developed by Russian defense manufacturer GosNIPP. Its mobile launcher is reportedly distinct from but "closely resembles" the INF-compliant Iskander-M TEL (9P78-1), which, if true, would complicate future arms control verification. Some analysts suggest the SSC-8 uses a 9P701 TEL that can carry four missiles.

MZKT-7930 Astrolog (Russian: M3KT-7930) 8×8 transporter-erector-launcher. Turbocharged diesel engine w/500 hp, a range of 1,000 km, and a max road speed of 70 km/h. Speed reduces to 40 km/h on field roads and 20 km/h off-road.

SSC-X-9 Skyfall (9M730 "Petrel")

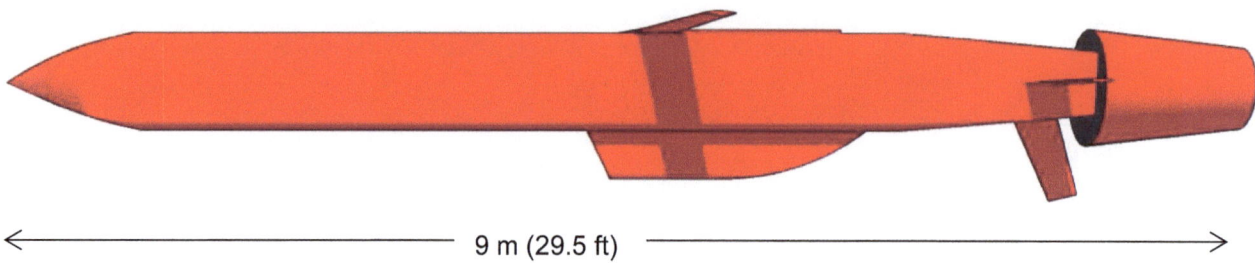

←――――――――― 9 m (29.5 ft) ―――――――――→

The SSC-X-9 (9M730 "Burevestnik", Russian: Буревестник; Petrel which literally means "storm crier" or "stormbringer", NATO reporting name: SSC-X-9 "Skyfall") is an experimental nuclear-powered, nuclear-armed cruise missile under development for the Russian Armed Forces. The missile is 9 m (29.5 ft) in length and claimed to have virtually unlimited range. Lot of controversy about this missile, especially the loss of life by 5 scientists during testing. The test also resulted in the release of radioactive particles. Recent photos show the missile with two booster rockets attached at the sides. A production model would probably require a booster at the tail so it would fit in the canister mounted on a transport-erector-launcher. In my opinion, it's a Rube Goldberg attempt that will probably never fly.

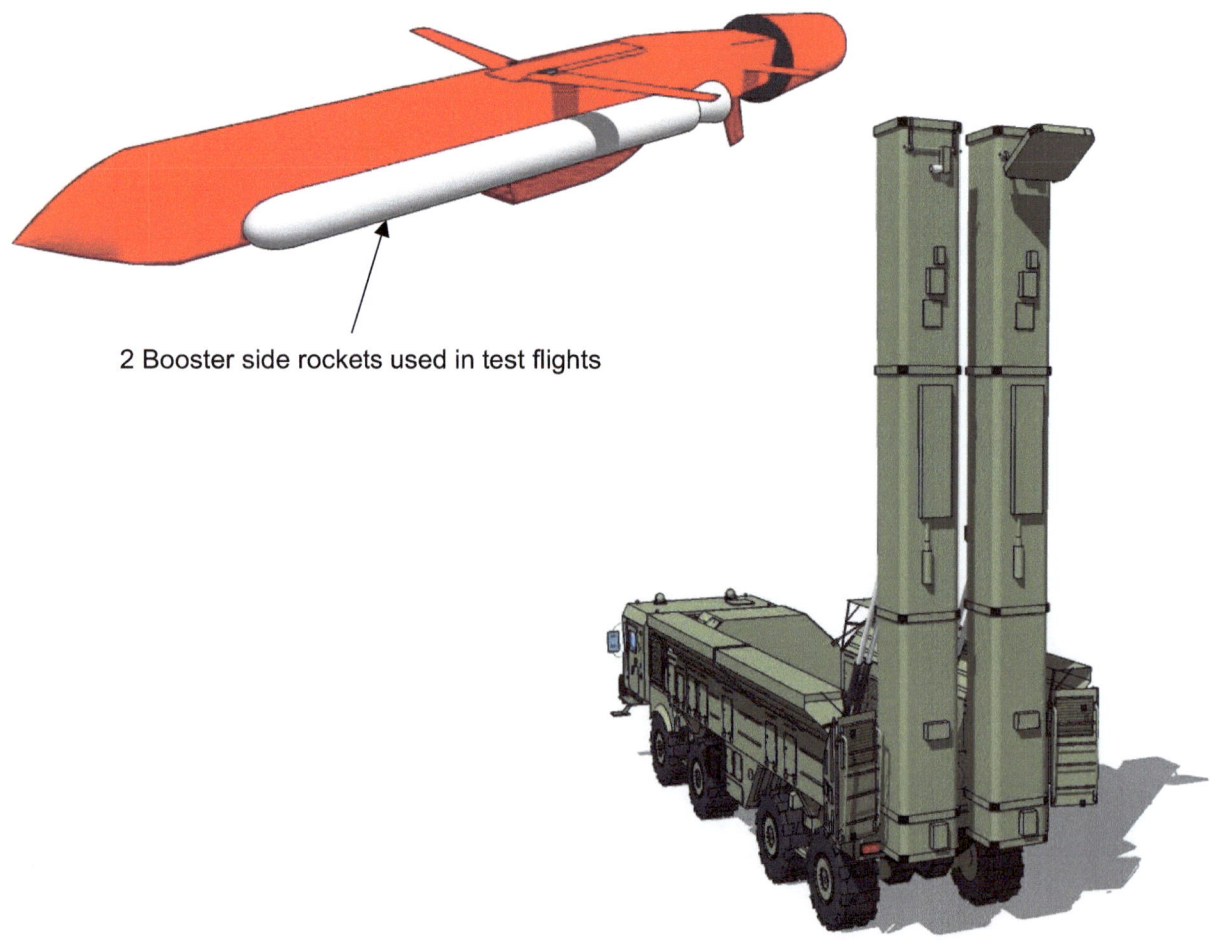

2 Booster side rockets used in test flights

SS-N-19 Shipwreck (P-700 "Granit" ASCM)

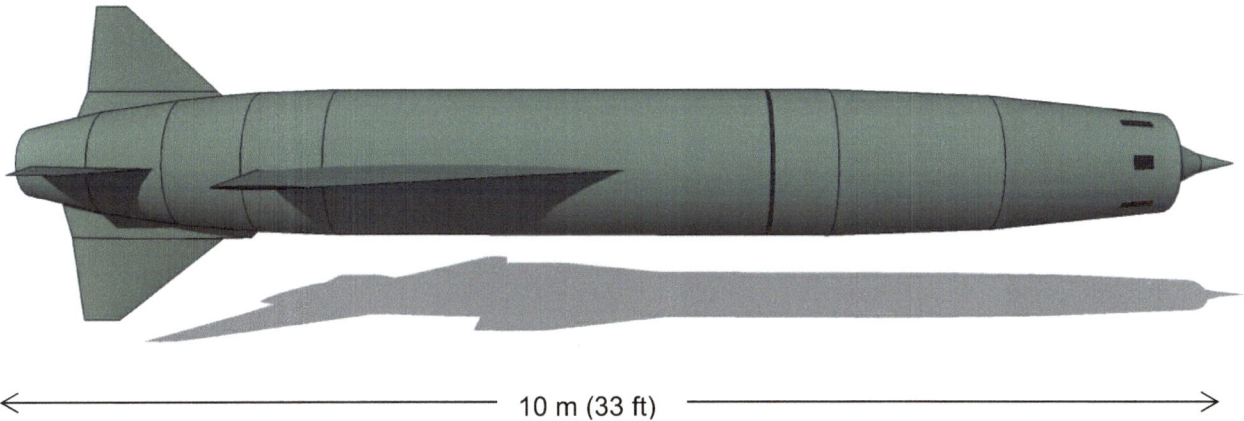

←———————— 10 m (33 ft) ————————→

The SS-N-19 Shipwreck (P-700 "Granit", Russian: П-700 "Гранит"; English: granite, NATO reporting name SS-N-19 "Shipwreck", GRAU designation is 3M45) is a naval anti-ship cruise missile the uses a ramjet. A stubby cylindrical solid-fuel rocket is fitted to the rear for launch; this booster stage is released when the missile enters sustained flight. The missile can be fitted with either a 750 kg HE warhead, a FAE warhead, or a 500 kt thermonuclear warhead. Range estimates vary between 400 km to 625 km. The guidance system is mixed-mode, with inertial guidance, terminal active radar homing guidance and also anti-radar homing. Mid-course correction is probable. It comes in surface-to-surface and submarine-launched variants, and can also be used against ground targets. The missile, when fired in a swarm (group of 4–8) has a unique guidance mode. One of the weapons climbs to a higher altitude and designates targets while the others attack. The missile responsible for target designation climbs in short pop-ups, so as to be harder to intercept. The missiles are linked by data connections, forming a network. If the designating missile is destroyed the next missile will rise to assume its purpose. Missiles are able to differentiate targets, detect groups and prioritize targets automatically using information gathered during flight and types of ships and battle formations pre-programmed in an onboard computer. They will attack targets in order of priority, highest to lowest: after destroying the first target, any remaining missiles will attack the next prioritized target.

Granit launchers on aircraft carrier Admiral Kuznetsov

SS-N-21 Sampson (RK-55 "Granat")

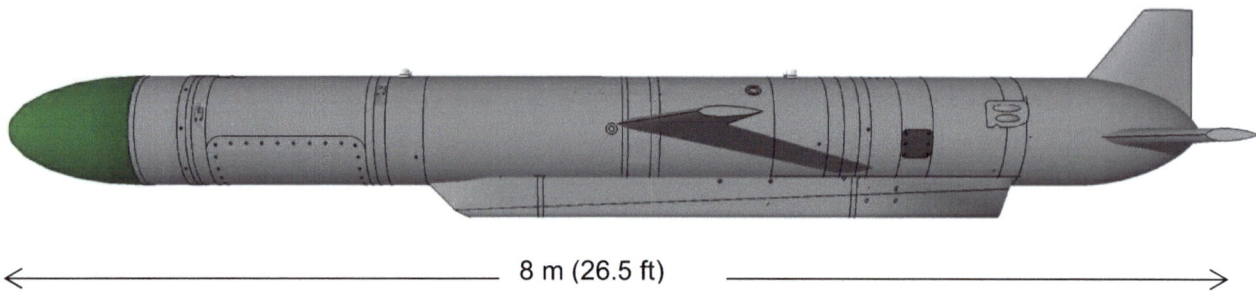

<- 8 m (26.5 ft) ->

The SS-N-21 Sampson (RK-55 "Granat", NATO: SS-N-21 "Sampson"; GRAU: 3K10) is launched from submarines through 533 mm torpedo tubes, and carries a conventional warhead. The missile has solid-propellant rocket booster and a R-95-300 or 36MT-37 turbofan. It has a length of 8 m (26.5 ft) and a wingspan of 3 m (10 ft). It has a range of 3,000 km (1,864 mi) at a speed of 720 km/h (447.4 mph). The ground-launched variant, the SSC-X-4 "Slingshot" was dismantled and destroyed due to the signing of the 1987 INF Treaty. The Russian Federation was reported to have deployed the derivative SS-CX-7/SS-CX-8 systems on February 14, 2017. The missile is very similar to the air-launched variant (AS-15 "Kent") but the Kent has a drop-down turbofan engine. Both have formed the basis of post-Cold-War missiles, in particular the 3M-54 Kalibr which has a supersonic approach phase.

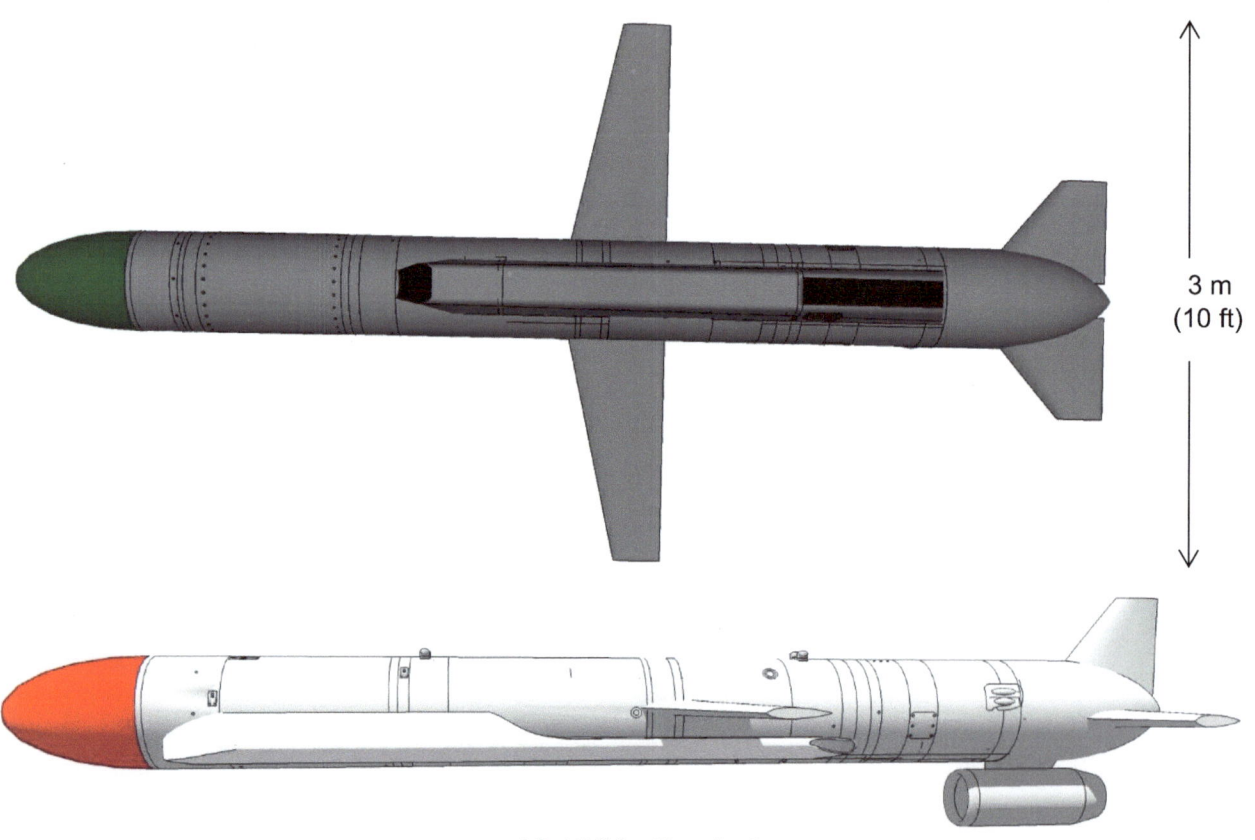

AS-15 "Kent" variant

SS-N-25 Switchblade (Kh-35 "Uran")

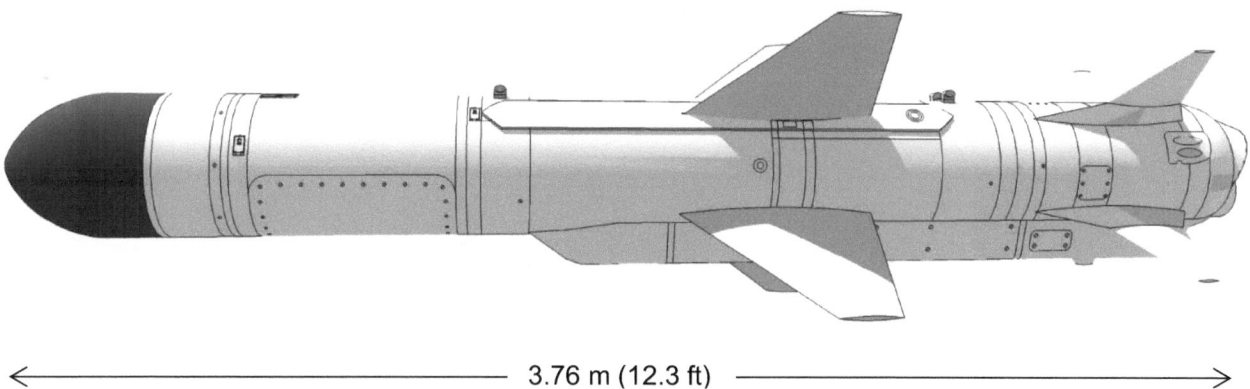

<- 3.76 m (12.3 ft) ->

The SS-N-25 Switchblade (Russian: X-35; "Uran", NATO designation "Switchblade", GRAU 3M24) is a turbojet subsonic anti-ship cruise missile with a range of 300 km (186 mi.). It has a solid-propellant rocket booster and a two-stage liquid-fueled propellent and has a cruise altitude of 5 to 15 m. It has an inertial guidance system with a Glonass Receiver and an active radar seeker and carries a conventional warhead of 480 kg. Russian officials claim that the KH-35U is immune to enemy countermeasures.

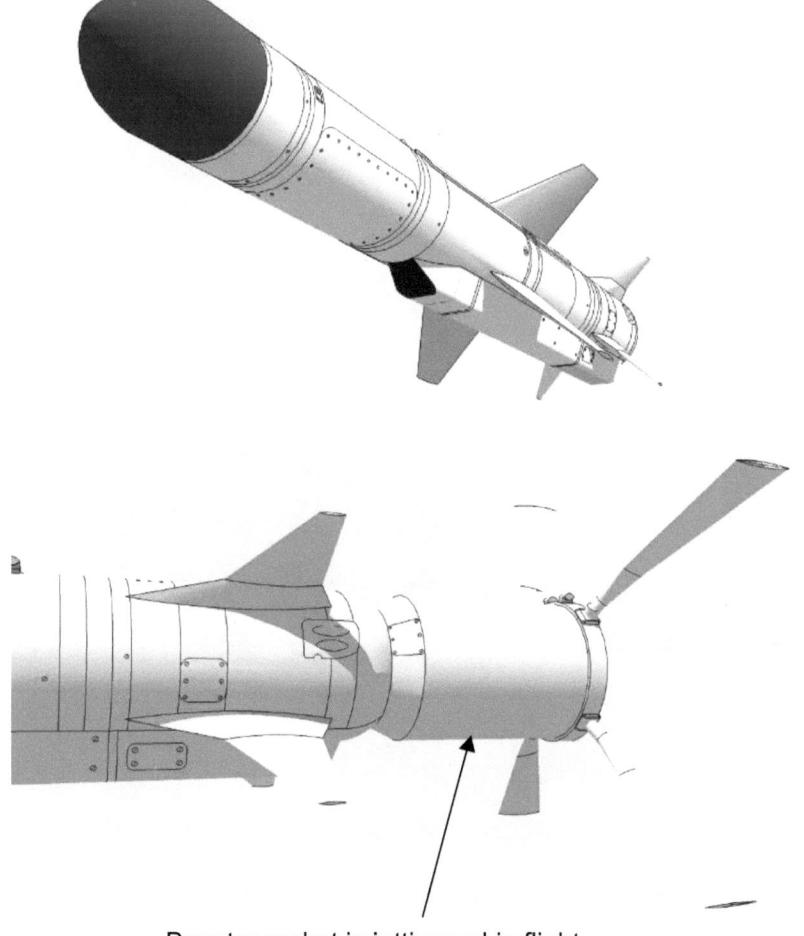

Booster rocket is jettisoned in flight

SS-N-26 Strobile (P-800 "Oniks" ASCM)
20 Launchers – 10 nuclear warheads

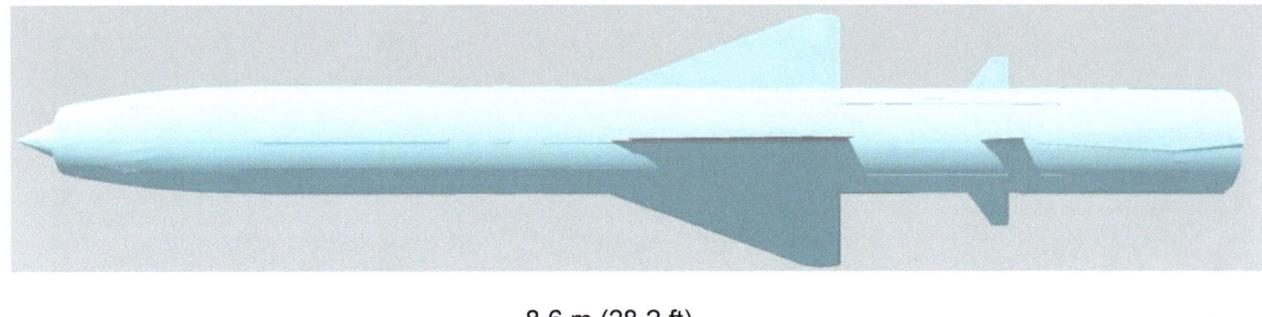

←──────── 8.6 m (28.2 ft) ────────→

The SS-N-26 Strobile (P-800 "Oniks", Russian: П-800 Оникс; English: Onyx, NATO designation SS-N-26 "Strobile", GRAU 3M55) also known in export markets as Yakhont (Russian: Яхонт; English: ruby), is a supersonic anti-ship cruise missile developed by NPO Mashinostroyeniya as a ramjet version of P-80 Zubr. The missile is launched from a 3K55 Bastion TEL and has a length of 8.6 m (28.2 ft) and a diameter of 0.7 m (2.3 ft). It can also be launched from a ship, and the air launched variant is the Kh-61. The missile has both an active and a passive inertial navigation system. It has a range of 600 km (370 mi.) and a speed of Mach 2. It has a thermonuclear warhead or 300 kg semi-armor piercing HE warhead with a delayed fuze. It is reportedly a replacement of the P-270 Moskit, but possibly also of the P-700 Granit. The P-800 was used as the basis for the joint Russian-Indian supersonic missile BrahMos.

3K55 Coastal defense Bastion transporter-erector-launcher

SS-N-27 Sizzler (3M-54K "Kalibr" AShM)

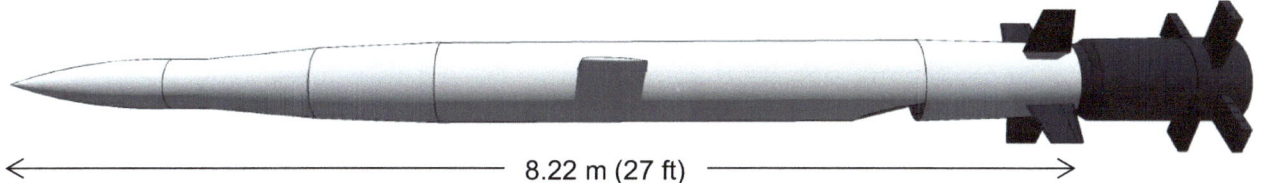

← 8.22 m (27 ft) →

There are many variants of the Kalibr missile, however I find the 3M-54K the most interesting.

The SS-N-27 Sizzler (3M-54K "Kalibr", Russian: Калибр, meaning caliber, also referred to as 3M14 "Biryuza", Russian: Бирюза, meaning turqoise, NATO designation: SS-N-27 "Sizzler" and SS-N-30A), is a submarine-launched (Akula, Yasen, Schula-B, and Lada class) anti-ship (AShM) and anti-submarine cruise missile. Other launch platforms include: naval ships, shipping containers, airplanes, and TEL. The missile has a length of 8.22 m (27.0 ft), with a 200 kg (440 lbs.) warhead. Its range is 440–660 km (270–410 mi). The missile is launched under the power of a solid-propellant rocket booster fitted with four small lattice stabilizers that is jettisoned when the missile reaches an altitude of 150 m. Once it has reached flying speed, the missile is powered by a small turbojet engine. It is a sea-skimmer with a final stage flight altitude of 4.6 meters (15 ft). Around 40 miles to its target, the forward section of the missile separates and ignites a solid booster, which rockets the missile to a supersonic speed of Mach 2.9 (3,550 km/h; 2,210 mph) in a zigzagging terminal run, reducing the time that target's defense systems have to react. During the terminal phase, the missile has a relatively high infrared signature. Kalibr-M is a new version under development with a larger warhead and an extended range of 4,500 km.

Sizzler shipping container and transport-erector-launcher

Shipping Container Launcher

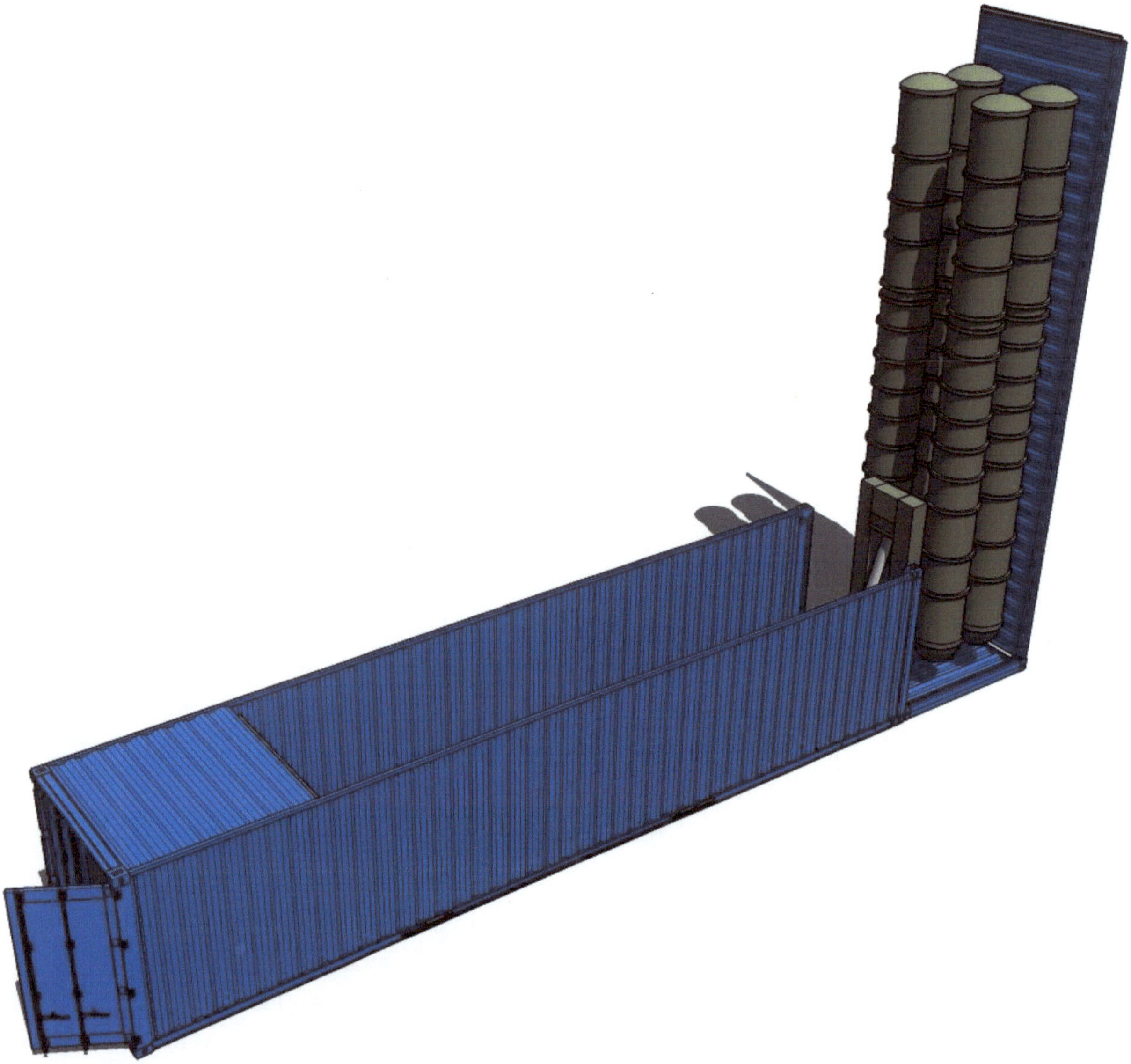

SS-N-30A (Land Attack Cruise Missile)

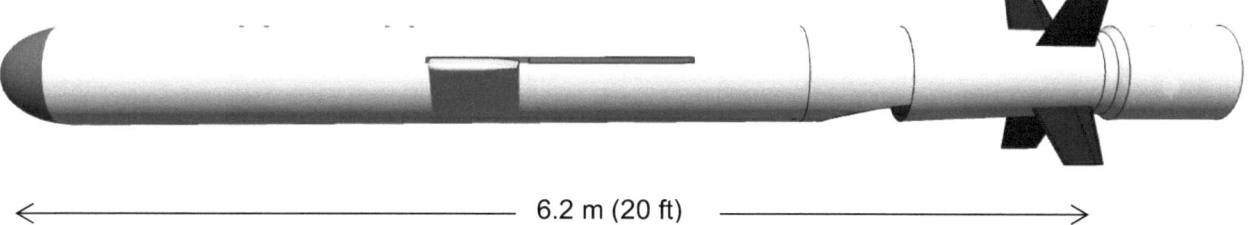

←──────── 6.2 m (20 ft) ────────→

A variant of the Kalibr Cruise Missile.

The SS-N-30 (A3M14K) is a land attack variant deployed by the Russian Navy. The submarine-launched weapon has a basic length of 6.2 m (20 ft), with a 450 kg (990 lb) warhead. Its range is 2,500 km (1,600 mi), allowing the Russian Navy to strike targets throughout Central/Western Europe from beyond the **GIUK** gap at subsonic speed of Mach 0.8. The missile is believed to fly 19.5 m (64 ft) above the sea and 50 m (164 ft) above the ground at speeds up to 965 km/hour. For most of the flight to the target area, the missile flies autonomously, following the pre-programmed route and turning points. Once over land, it uses a terrain-following flight path that will make it a difficult target for enemy air defenses. Low-level flight mode poses a higher load on the wings and missile structure than flight over the sea, so the land-attack missile has slightly redesigned wings of shorter span and deeper chord, plus a stronger structure.

GIUK Gap (Greenland, Iceland, United Kingdom)

AS-15 Kent (KH-55 ALCM)

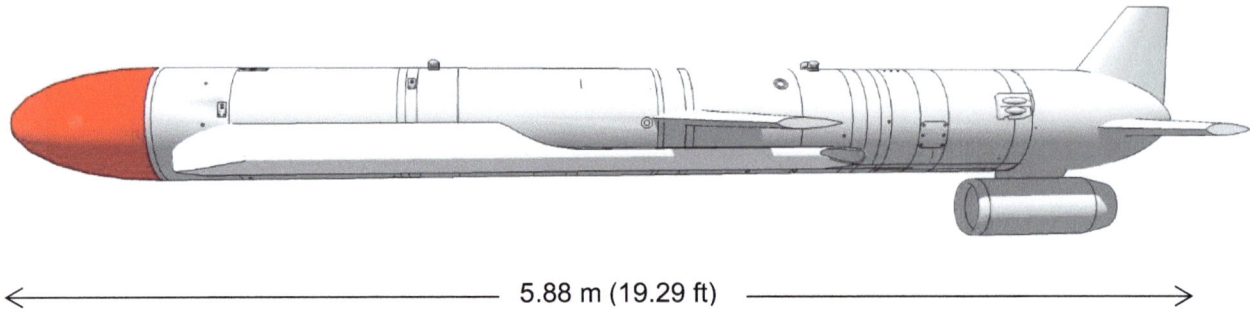

←——————— 5.88 m (19.29 ft) ———————→

The AS-15 Kent (Russian: KH-55, also known as X-55 and RKV-500; NATO reporting name: AS-15 "Kent") is a Soviet/Russian subsonic air-launched cruise missile, designed by MKB Raduga. It has a range of up to 2,500 km (1,553 mi.) and can carry nuclear warheads. The AS-15 is launched exclusively from bomber aircraft and has spawned a number of conventionally armed variants mainly for tactical use, such as the Kh-65SE and Kh-SD, but only the Kh-101 and Kh-555 appear to have made it into service. It can be launched from both high and low altitudes, and flies at subsonic speeds (Mach 0.75) at low levels (under 110 m/300 ft altitude). It is guided through a combination of an inertial guidance system plus a terrain contour-matching guidance system which uses radar and images stored in the memory of an onboard computer to find its target. This allows the missile to guide itself to the target with a high degree of accuracy. Having reached the area where the target is located, the Kh-65 rises to a higher altitude and its active-radar target seeking system turns on. The original Kh-55 has a drop-down engine; the Kh-65SE has a fixed external turbojet engine, while the Kh-SD had its engine inside the body of the missile. Current-production versions are equipped with the increased power of 450 kgf Russian-made NPO Saturn TRDD-50A engine. The AS-15 is very similar to the RK-55 Relief (SS-N-21 "Sampson"), but the AS-15 has a drop-down turbofan engine.

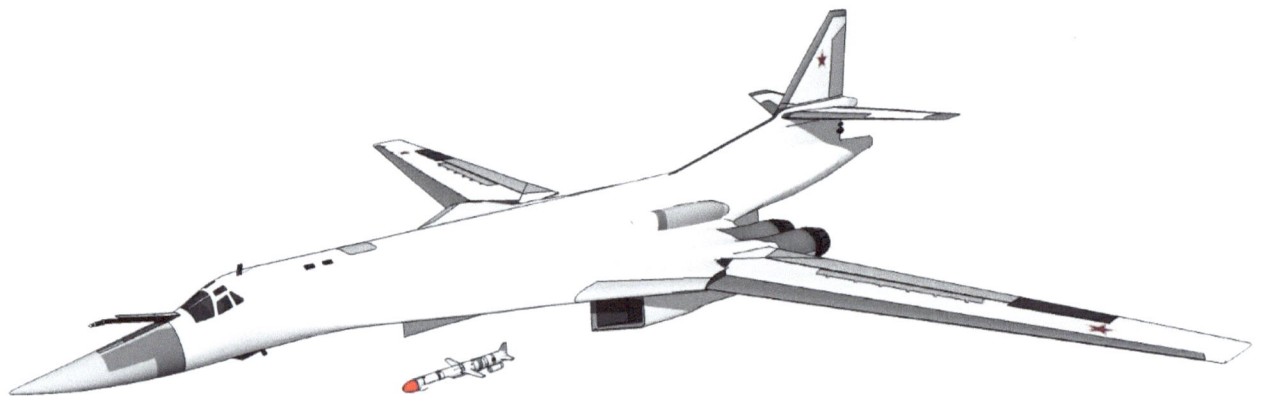

Tu-160 Blackjack releases AS-15

AS-23A / AS-23B Kodiak (Kh-101/Kh-102 ASCM)

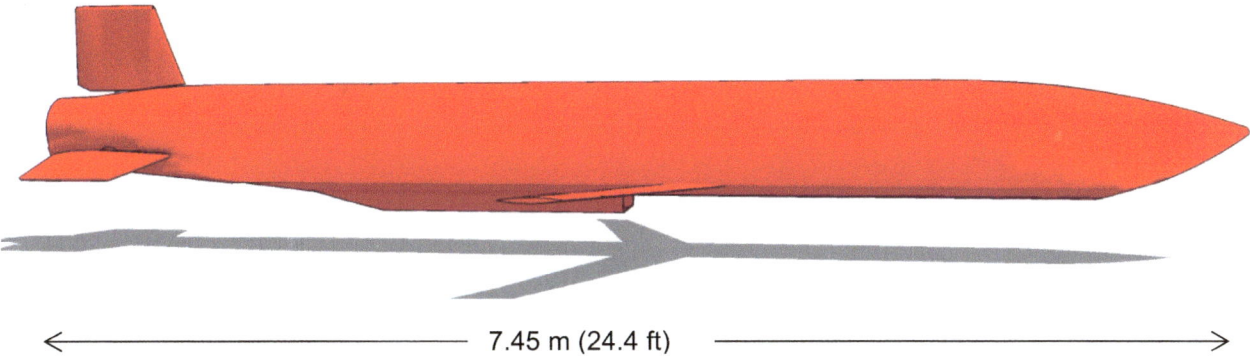

←―――――――――― 7.45 m (24.4 ft) ――――――――――→

The AS-23A and AS-23B Kodiak (Kh-101 and Kh-102) are long range subsonic air-launched cruise missiles. The AS-23A (Kh-101) carries a conventional warhead, while the AS-23B (Kh-102) is believed to carry a nuclear payload. This is the latest development of the Kh-55, incorporating a low radar cross-section of about 0.01 square meters. The missile is specifically designed for air-launch, abandoning the circular fuselage cross-section of the Kh-55 for a nose and forward fuselage section "aerodynamically shaped" to produce lift. It is 7.45 m (24.4 ft) long with a wingspan of ~ 3 m (~10 ft) with a launch weight of 2,200–2,400 kg (4,900–5,300 lb) and is equipped with a 400 kg (880 lb) high-explosive, penetrating, or cluster warhead, or a 250 kT (or 450 kT) nuclear warhead for the Kh-102. The missile is powered by a TRDD-50A turbojet producing 450 kg (990 lb) of thrust to cruise at 700–720 km/h (430–450 mph; Mach 0.57–Mach 0.59) with a maximum speed of 970 km/h (600 mph; Mach 0.79) while flying 30–70 m (100–230 ft) above the ground, and hit fixed targets using a pre-downloaded digital map for terrain following and GLONASS/INS for trajectory correction to achieve accuracy of 6–10 meters; it is claimed to be able to hit small moving targets like vehicles using a terminal electro-optical sensor or imaging infrared system. Range estimates vary from >2,000 km (1,200 mi), to 4,500–5,000–5,500 km (2,800–3,100–3,400 mi), to as much as 10,000 km (6,200 mi) with a flight endurance of 10 hours; long range is essential since Russia has few bases abroad and cannot provide distant fighter escorts. The Tu-95MS "Bear H" can carry eight of the weapons on four under-wing pylons (the missile will not fit internally) and the Tu-160 can be outfitted with two drum launchers each loaded with six missiles for 12 total, but the smaller Tu-22M3 will continue to carry the Kh-555, although it can also carry the Kh-101/Kh-102. It can also be affixed to the Su-34 fighter jet. The missiles have an onboard Electronic Warfare (EW) defense system. Long-range ALCMs missiles are considered less of a threat to global stability because they cannot execute short-notice, disarming attacks like ground-based ballistic missiles. Future air-launched cruise-missile development will focus on extending the range (the Kh-BD Bolshoi Dalnosti, meaning 'long-range' in Russian).

Tu-95MSM with 8 AS-23 ALCM Photo credit: Dmitry Terekhov

Russian Tactical Nuclear Weapons

Recent press reports state that Russia is moving tactical nuclear weapons or nuclear custodial units along with their delivery vehicles to Ukraine. It has been suggested that Putin is signaling the threat of tactical nuclear weapons to forestall recapture of annexed Ukrainian territories. ("descalation by escalation"). This is consistent with Putin's remark in October 2018, "our concept is based on a reciprocal counter strike … This means that we are prepared and will use nuclear weapons only when we know for certain that some potential aggressor is attacking Russia, our territory". Apparently, Russia has lowered its threshold for first use of nuclear weapons.

According to published reports, Russia has 1,912 nonstrategic (tactical) nuclear warheads. Russian weapons delivery vehicles deployed near Ukraine are dual-capable, meaning that they can launch either conventional or nuclear weapons. The biggest user of tactical nuclear weapons in the military is the **Navy**, which is estimated to have roughly **935 warheads** for use by land-attack cruise missiles, anti-ship cruise missiles, anti-submarine rockets, anti-aircraft missiles, torpedoes, and depth charges. The **Air Force** is the second-largest user of tactical nuclear weapons, with roughly **500 warheads** assigned for delivery by Tu-22M3 (Backfire) intermediate-range bombers, Su-24M (Fencer-D) fighter-bombers, the Su-34 (Fullback) fighter bomber, and the MiG-31K.

Tactical Nuclear Missiles

Russian Defense Minister Sergei Shoigu announced in December 2019 that the upgrade of all **Army** missile brigades to the 350-kilometer (217 mile) range **SS-26** (Iskander) short-range ballistic missile had been completed. This includes at least 12 brigades: four in the Western Military District; two in the Southern Military District. Each brigade has 12 launchers, each with two missiles for a total of 24 missiles (at least one reload is in storage). In 2019, Izvestia quoted unnamed defense ministry sources saying that each brigade would receive an additional battalion so that each brigade in the future would have 16 launchers with 32 missiles. It has been estimated that there are roughly **70 warheads** for short-range ballistic missiles. There are also unconfirmed rumors that the **SSC-7** (9M728 or R-500) ground-launched cruise missile <u>may</u> have nuclear capability.

Russia developed and deployed a dual-capable ground-launched cruise missile (GLCM) identified as the 9M729 (**SSC-8**) with a range of 2,500 km. Its development prompted the 2019 U.S. withdrawal from the 1987 INF Treaty. Russia initially tested the 9M729 to prohibited ranges from a fixed launcher, then tested it from a mobile launcher. Russia has deployed four battalions in the Western, Southern, Central, and Eastern Military Districts with nearly 100 missiles (including spares).

Russian 9M728 cruise missiles, left, and 9M723 short-range ballistic missiles, right.

It is unknown if Russia has added 9M729 battalions beyond the four reported in December 2018. The 9M729 has probably been added to a fifth brigade: the 26th Missile Brigade outside

Luga about 125 kilometers (about 78 miles) south of St. Petersburg. The Russian military planned to add a fourth battalion to each existing Iskander brigade, each of which previously comprised three battalions of four launchers each (each launcher carries two missiles and probably two reloads). It remains to be seen if this means that 9M729 launchers will be added to all of Russia's 12 Iskander brigades. Some analysts suspect the SSC-8 uses the 9P701 TEL. As of December 2018, Russia had produced fewer than 100 SSC-8 missiles.

Nuclear Artillery

1 Kiloton

2S7 Pion Self-Propelled Gun

3BV2 Kleshchevina Nuclear Shell

The **2S7 Pion** ("peony"; NATO designation: M-1975) or **Malka** is a Soviet self-propelled 203mm cannon capable of firing the **3BV2 Kleshchevina** nuclear shell with a range of 11 to 18 miles. The 2S7 features the massive 2A44 howitzer with a length of 56 calibers. More than 250 2S7 Pions were built. The 3BV2 Kleshchevina nuclear shell was introduced in 1977. Derived from the 3BV3 designed for 152mm guns. It is a Plutonium-based weapon with a yield of about **0.5 to 1 Kt**. It can be fired from the 2S7 Pion or the towed B-4M.

2.5 Kiloton

The **2S19 Msta** (Russian: Мста) is a 152.4 mm self-propelled howitzer capable of firing the 3BV3 nuclear shell with a range of 15 to 18 miles. The 3BV3 is a Plutonium-based weapon with a yield of about 2.5 Kt. It takes about 30 minutes from opening the loading container to the firing command. This projectile was last produced in 1990. Since the projectile has a lifespan of 10 years, all its nuclear charges may have been dismantled and destroyed in 2000. Nuclear artillery has been almost entirely replaced by mobile tactical ballistic missiles with nuclear warheads.

Nuclear Blast Yield

Below is a simulation of a 1 Kt blast on Luhansk.

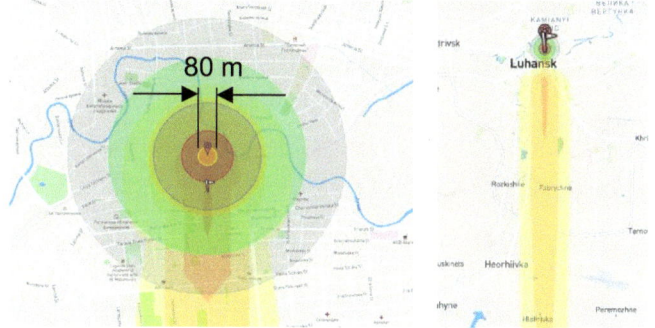

Estimated fatalities: 5,560
Estimated injuries: 11,290
Radius
Fireball radius: 80 m (0.02 km²)
Heavy blast damage (20 psi): 220 m (0.15 km²)
Moderate blast damage (5 psi): 460 m (0.66 km²)
Thermal radiation (3rd degree burns): 0.5 km (0.79 km²)
Radiation radius (500 rem): 0.84 km (2.2 km²)
Light blast damage radius (1 psi): 1.18 km (4.35 km²)

Below is a simulation of a 2.5 Kt blast on Luhansk.

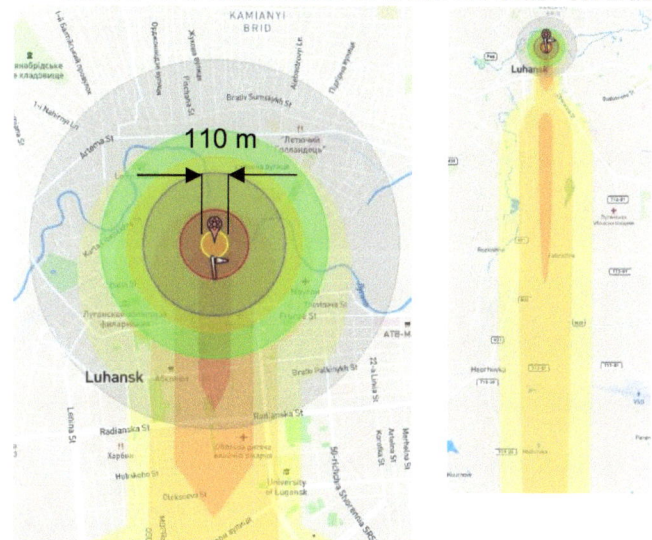

Estimated fatalities: 9,010
Estimated injuries: 20,800
Radius
Fireball: 110 m (0.04 km²)
Heavy blast damage (20 psi): 300 m (0.27 km²)
Moderate blast damage (5 psi): 0.62 km (1.21 km²)
Thermal radiation (3rd degree burns): 0.76 km (1.81 km²)
Radiation (500 rem): 0.98 km (3.03 km²)
Light blast damage (1 psi): 1.6 km (8.01 km²)

www.ingramcontent.com/pod-product-compliance
Lightning Source LLC
Chambersburg PA
CBHW051216220526
45473CB00003B/1052